SHARK

SHARK

THE ILLUSTRATED BIOGRAPHY

DANIEL C. ABEL & SOPHIE A. MAYCOCK

Illustrations by Adam Hook

Princeton University Press
Princeton and Oxford

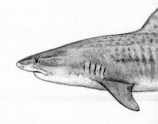

CONTENTS

Published by Princeton University Press
41 William Street, Princeton, New Jersey 08540
99 Banbury Road, Oxford OX2 6JX
press.princeton.edu

Library of Congress Control Number 2024944556

ISBN 978-0-691-26167-6

Ebook ISBN 978-0-691-26166-9

Typeset in Trajan and Caslon

Printed and bound in Malaysia
1 3 5 7 9 10 8 6 4 2

British Library Cataloging-in-Publication
Data is available

This book was conceived, designed, and produced by
UniPress Books Limited
Publisher: Jason Hook
Project manager: Richard Webb
Art direction: Alexandre Coco
Design: Paul Palmer-Edwards
Illustrator: Adam Hook
Infographics: Rob Brandt

Cover design: Paul Palmer-Edwards
Cover images: Adam Hook

Sharks have everything a scientist dreams of. They're beautiful—God, how beautiful they are! They're like an impossibly perfect piece of machinery. They're as graceful as any bird. They're as mysterious as any animal on Earth. No one knows for sure how long they live or what impulses —except for hunger—they respond to.

PETER BENCHLEY

PREFACE

Books about sharks abound, but *Shark: The Illustrated Biography* represents a variation on the typical shark book theme. It is not, as you might expect, a conventional biography, which is an account of an individual's life told by another. This book is an unauthorized biography, since obtaining permissions, and hence cooperation, from the subjects would be a bit of a bridge too far. But like any biography it tells the story of our subjects' whole lives, from their birthplace and childhood through to their family history and relationships, detailing their talents and skills, significant events and life challenges, and, ultimately, their death.

The most readable and enjoyable biographies transcend dry, chronological listings of the events of a subject's life; they are a continuous story that enriches these events with color and context, highs and lows, failures and triumphs. While this book absolutely will include facts, they are presented with a minimum of scientific jargon, unnecessary detail, or out-of-context factlets. Additionally, this *Illustrated Biography* will include stories about sharks, especially the love–hate relationship we have had with sharks throughout our shared history.

While all sharks share the characteristics that define them as sharks and not, say, grouper or tuna, there is so much variety among sharks that telling their story cannot be the biography of a single species. Thus, instead of chronicling the life of an individual shark, or even an individual species, this biography focuses on sharks as a group through the eyes (so to speak) of four species—the Great White Shark, Sandbar Shark, Spiny Dogfish, and Smallspotted Catshark—which have been selected for more detailed treatment throughout. As you will see a little later, these are not typical sharks. They are, however, the sharks that populate media accounts and documentaries, are displayed in large aquaria, and may even be found on your plate, and thus are those with which you might be at least somewhat familiar. Much of what applies to the four selected species also applies to their close relatives, and these cousins play roles throughout the book as well.

Currently, about 550 species of sharks are known to science, with numerous others certainly awaiting discovery (we discuss this in depth in Chapter 3).

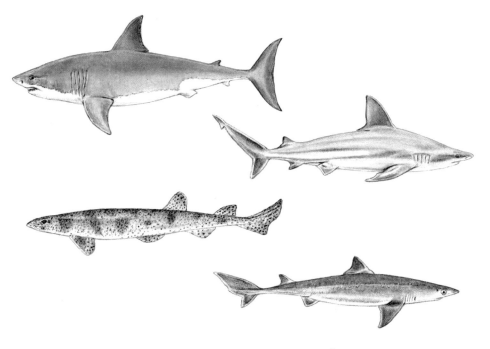

THE MAIN CHARACTERS
*Although this book covers all of sharkdom, we focus on these four species
(top to bottom): White Shark, Sandbar Shark, Smallspotted Catshark,
and Spiny Dogfish.*

There are, therefore, so many additional stories to tell beyond the four species on which we focus. Since writing a 550-volume *Biographical Sketches of the Sharks* cannot happen at the moment, some of the more interesting stories from the species of sharks not highlighted will also find a home in these pages, as will accounts from the catalog of the rays, sharks' closest living relatives.

A few words on how this book is organized are called for. Chapter 1, Sharks: An Introduction, defines what makes a shark a shark, introduces you to the four species we highlight, and sets the stage for the rest of the biography. Chapter 2, Roots, tells the fascinating evolutionary story of sharks, explaining when they first appeared and how they changed over time to become the highly efficient and effective predators of today. Chapter 3, The Shark Family Tree, tells the story of how sharks and rays are classified, and explores the biodiversity of the group. Chapter 4, Sharks in their Habitats, could also have been called Shark

Ecology. This brings us to a series of three chapters that most closely conform to the structure of a biography: prenatal life, early life, and adult life. These are Chapter 5, Sharks in the Womb, Chapter 6, The Wonder Years, and Chapter 7, The Adult Years. Chapter 8, Sharks That Break the Rules, was one of the most fun to write. Chapter 9, A Tale as Old as Time, is a short cultural biography of sharks, specifically their relationship to humans. The book concludes with a sobering but hopeful chapter, The Final Chapter?, which considers the threats we have posed to sharks—both historically and in the present day—as well as the important scientific and conservation work being conducted by the public, nonprofit organizations, and the shark scientific research community.

Shark: The Illustrated Biography celebrates the stunning elegance and charisma of sharks, their essential ecological roles, their astounding adaptations, their surprising behaviors, and their biodiversity, as well as their enduring place in our lives and imaginations. And if that is not enough reason to write a biography of sharks, consider this Douglas Adams quotation: "There were sharks before there were dinosaurs, and the reason sharks are still in the ocean is that nothing is better at being a shark than a shark."

1

SHARKS

AN INTRODUCTION

THE GIFT OF THE PRESENT

Imagine you lived in a world where anything was possible, where you were constrained only by the limits of your imagination. What if you could go anywhere in the world or even back in time in the blink of an eye? Where would you choose to go and what would you like to see?

If you are a shark enthusiast (which is presumably the case because you picked up this book), maybe you would remain in the present and swim with the Greenland Shark (*Somniosus microcephalus*), wandering lazily in the frosty waters of the Arctic, amid a labyrinth of colossal, blue-tinged icebergs. Or perhaps you would prefer to hang out with a Caribbean Reef Shark (*Carcharhinus perezi*), meandering among magnificent corals which glisten in every color imaginable, immersed in tropical waters that teem with a wondrous assemblage of smaller fish, all potential prey for this magnificent reef shark. Would you instead choose to explore the endless expanses of the open ocean, a world distant from land, or down, *way down*, in the mysterious, cavernous depths of the deep sea, where sharks unlike any you have ever seen loom out of absolute darkness?

Imagine you could take a trip back in time. Maybe you would like to visit the giant shark called Megalodon (*Otodus megalodon*)—a brute of a creature that human eyes have never marveled at, as it lived millions of years before we even evolved. No fish in today's ocean reaches the size of this mighty creature, although the plankton-eating Whale Sharks (*Rhincodon typus*) and Basking Sharks (*Cetorhinus maximus*) do come close. Perhaps you might like to visit an era about 420 million years ago (MYA), when sharks were the newest kids on the block, evolutionarily speaking. Alternatively, maybe you'd secure a ticket to the more recent Carboniferous Period, about 360 to 300 MYA, an era with a sobriquet that would make any shark lover drool: the Golden Age of Sharks. In the huge diversity of marine and freshwater habitats of this time, there were many more kinds of sharks than exist today.

Maybe you would like to transpose yourself into the life of the first human who ever came face to face with a shark, as many as 300,000 years ago. What on earth must our early ancestors have thought? How did they feel? Your curiosity might also make you want to see into the future, to investigate how the major environmental impacts of the human era, the Anthropocene, have eventually played out. Will sharks have outlived humans, or vice versa? Or did we find a way to prosper together?

Given the absence of the level of techno-wizardry required to design and build a time machine, or the physical impossibility of doing so, we must, alas, be content with living in the present. But what a gift the present is—from the perspective of shark lovers, that is. How fortunate that we need do no more than visit the shoreline for a chance of seeing the mesmerizing spectacle of a Sandbar Shark (*Carcharhinus plumbeus*) surfing through the curl of a breaking wave, or a Blacktip Shark (*Carcharhinus limbatus*) corkscrewing out of the water, perhaps to dislodge a parasite, to assert some sort of dominance, or simply to express the joy of being a Blacktip Shark.

UNDERWATER SURFER
A shark rides inside the curl of a wave as it patrols the beach—grace personified.

There's plenty of fish in the sea

We are lucky enough to live on a planet boasting more than 550 different species of sharks, with dozens of new species likely remaining to be discovered. There are sharks practically anywhere with enough water to contain them. They can be found cruising in the pelagic realm (in open ocean, far away from land), in the expansive zone above the ocean floor, along our coastlines in ecosystems that include estuaries, reefs, and mangroves, and even in freshwater lakes and rivers. Sharks range from pole to pole and from the shallows to the ocean depths, living in every habitat imaginable. Some undergo incredible migrations, traveling almost around the world and back, whereas others live in the same small cove for their entire lives. Some sharks hunt within dense shoals of fish, some dive down to gnash at squid, while others chase down marine mammals like seals and dolphins. Many sharks prefer to snack on other sharks and some are even cannibals.

In fact, sharks themselves vary just as much as the habitats in which they live. Whereas some sharks—the Whale Shark, for example—are longer than a double-decker bus, others, like the Dwarf Lanternshark (*Etmopterus perryi*), would sit comfortably in the palm of your hand. Coming in grays, black, browns, yellows, oranges, and even blue, such as the appropriately named Blue Shark (*Prionace glauca*), sharks may sport fetching spots, tiger stripes, or countershading (dark on top, light on bottom). Whale Sharks even have a combination of all these patterns. Some sharks, like the Tope (*Galeorhinus galeus*), which is also called the Soupfin Shark, have sharp, pointed teeth, whereas others, like Nurse Sharks (*Ginglymostoma cirratum*), have flattened, platelike teeth. Some species, Great White Sharks (*Carcharodon carcharias*), for example, may even change their tooth shape completely as they age. Horn Sharks (*Heterodontus francisci*) have two types of teeth in their mouth at once, with cusped teeth up front and molar-like teeth in the back, and they possess enlarged jaw muscles for powering the teeth to crush snails, urchins, and crabs.

Where one species may have a stereotypically streamlined body rippling with muscle, such as the reef sharks (*Carcharhinus* species), others, like the Small-spotted Catshark (*Scyliorhinus canicula*), have a slimmer, more elongated, almost

AERIAL SHARK

Few images are as striking as a White Shark escaping the watery bonds
of the sea in its acrobatic pursuit of a Cape Fur Seal.

eel-like body. Then there is Megamouth (*Megachasma pelagios*), a blob of a beast with a bulbous head. Not all sharks even have the same complement of fins. The dogfish (family Squalidae) have lost the anal fin and the cowsharks (family Hexanchidae) are missing a dorsal fin. The fins of some sharks—for example, the angel sharks (family Squatinidae)—have evolved to be more raylike, consistent with their evolutionary strategy to adopt a more benthic lifestyle (living near to the ocean floor). Finally, to complete this short venture into shark diversity, the sawsharks (order Pristiophoriformes) even have toothed saws sticking out from the front of their face. Imagine how tough it would be for a sawshark to swim.

* * *

Since we are incapable of whisking you through space and time, our next best option is to create words and images that do so. The goal of this biography is to transport you, within your imagination, into the life of a shark: to explore their worlds, learn how they live, and empathize with their experiences. What is it like to be a shark? When did sharks first appear? Where are they born? How do they live and for how long? It is quite a story! But to begin this journey, you first need to know what a shark actually is.

A SHARK'S TALE

Sharks are very different from tuna, swordfish, mullet, sea bass, sardines, and other types of fish. Yet, make no mistake, sharks are indeed fish, but fish that are much more closely related to skates and rays than to tuna or swordfish. In fact, sharks, skates, rays, and a relatively obscure group called chimaeras (or ghost sharks) are all grouped together in the taxonomic class Chondrichthyes, a word that derives from the ancient Greek for "cartilage" and "fish." Unlike

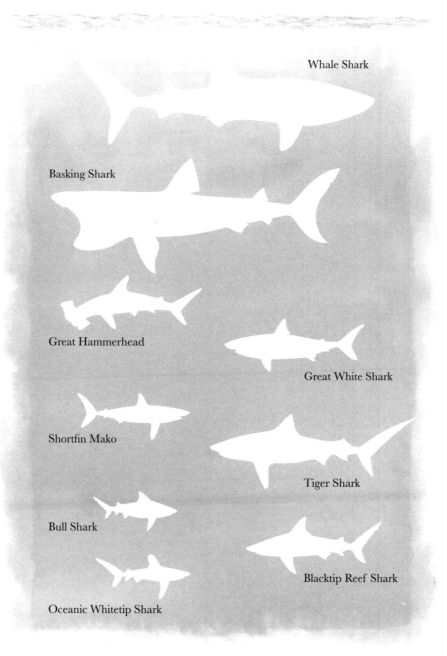

Whale Shark

Basking Shark

Great Hammerhead

Great White Shark

Shortfin Mako

Tiger Shark

Bull Shark

Blacktip Reef Shark

Oceanic Whitetip Shark

DIVERSITY OF SHARKS
With around 550 species, sharks represent a diverse assemblage,
including the largest fish in the sea, the Whale Shark.

other fish, which possess a skeleton of relatively dense bone, sharks, skates and rays, and chimaeras have a skeleton built from a lighter, more flexible material: cartilage. That single anatomical distinction is the main criterion for historically placing the more than 33,000 kinds of bony fish in the taxonomic class Osteichthyes (meaning "bony fish") and the approximate 1,300 species of sharks, skates, rays, and chimaeras in the class Chondrichthyes.

We humans also have cartilage in our body, multiple kinds, in fact: in our nose, ears, bronchial tubes, trachea, ribs, ends of the long bones, and the discs between our vertebrae. Cartilage makes up about 0.6 percent of our body weight, but as much as 6 to 8 percent of a shark's. Believe it or not, cartilage is actually less dense than honey (but it does not flow). Humans need hardened bone because a skeleton made completely from cartilage simply would not have the rigidity to support our bodies against the force of gravity. But it works fantastically for sharks in their more buoyant, watery world.

In spite of its lightness, cartilage gives sharks the level of structural support that bone gives us and serves as points of attachment for muscles and tendons, for swimming, eating, and so on. At the same time, the lightness of cartilage also helps to compensate for the heaviness of the muscle-bound chondrichthyan body. Most bony fish possess a gas-filled internal balloon called a swim bladder to help them remain buoyant, which both saves energy and allows them to exploit living spaces, such as the tight crannies around coral reefs, that would otherwise be inaccessible. Sharks and their other chondrichthyan cousins lack a swim bladder, which would have been a major obstacle to the evolutionary success of the group if cartilage had not come to the rescue, substituting a light skeleton in place of a heavy one and compensating in part for their brawny bulk.

When you see a shark, especially iconic species like Great White Sharks, Shortfin Makos (*Isurus oxyrinchus*), Great Hammerheads (*Sphyrna mokarran*), and Grey Reef Sharks (*Carcharhinus amblyrhynchos*), you immediately recognize it as a shark. What was it that informed you? Was it the imposing jaws filled with row upon row of impressive teeth? Was it the five to seven gill slits on either side of their body? Or maybe it was the asymmetrical caudal fin (or tail fin), with the upper lobe longer than the lower? These attributes are indeed distinguishing features of sharks. Yet sharks and their cousins have a number of additional defining characteristics, the most prominent of which is claspers, tubular extensions of the inner margin of the pelvic fins responsible for sperm transfer in males. We will go into more detail about these later, but for now all you need to know is that claspers make sharks an oddity among fish, as they allow for internal fertilization. Among bony fish, fewer than 3 percent of species use internal fertilization to breed (the mollies and guppies in aquaria, for example), whereas all sharks and other chondrichthyan fish fertilize internally. This adaptation has been of fundamental importance in the life history strategy of the group, in most cases specifically for producing a small number of relatively large offspring capable of surviving on their own from day one.

Another distinguishing feature of sharks and their close relatives is their ability to grow new teeth throughout their lives. Where many adult animals, including humans, have one set of permanent teeth, firmly embedded in bony sock-

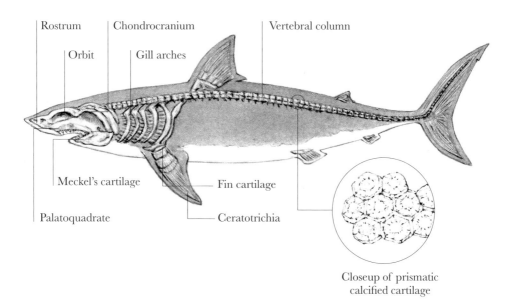

Rostrum

Orbit

Chondrocranium

Gill arches

Vertebral column

Meckel's cartilage

Palatoquadrate

Fin cartilage

Ceratotrichia

Closeup of prismatic
calcified cartilage

THE CARTILAGINOUS SKELETON OF SHARKS
*Cartilage evolved as a lighter and more flexible alternative to bone
and has been shown to be nearly as strong.*

ets in their jaws, sharks continue to bud new teeth deep in the fringes of their mouths all their lives. This is why you can see so many rows of teeth within a shark's mouth—a Great White Shark can have as many as 300 teeth at any one time and thousands during its lifetime! Not all the teeth are sharp and at the ready, but in most cases are in different stages of development, anchored on a loose assembly line of connective tissue rather than in cartilage. This polyphyodonty, or revolver, dentition, a condition also found in crocodiles, manatees, elephants, and kangaroos, means sharks will never lose or dull all their teeth and be unable to hunt, because if a tooth is shed at the front, another one simply pops up (so to speak) from behind to replace it. From an evolutionary perspective, always having one or more rows of functional teeth had to have signif-

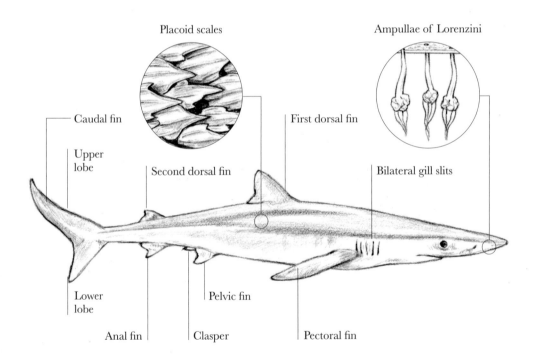

CHARACTERISTICS OF SHARKS

Sharks share some similarities with their bony fish cousins, but there are major differences in their skeletons, scales, senses, jaws, and even fins.

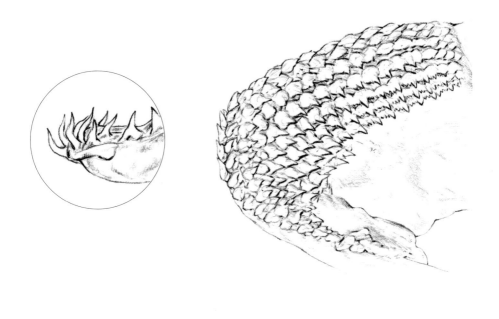

SERIAL TOOTH REPLACEMENT
*Another major departure from the typical vertebrate plan is polyphyodonty, having a
continuous supply of teeth for biting, shearing, and/or grasping, as well as for crushing.*

icant survival value to justify the mineral resources and energy required to
produce teeth and subsequently shed them. Terrestrial apex predators frequently lose or wear down their teeth, and these are not replaced, putting the
animal at risk of starvation. The shark dentition plan, however, means that
they will never lose all their teeth and be unable to hunt.

Sharks, skates, and rays are also distinguished by the structure of their skin.
While shark skin may look smooth, it is actually incredibly rough due to its dermal denticles (also called placoid scales), which consist of (in most sharks) up to
thousands of interlocking tiny plates. If you think about this carefully, you'll
notice that the name is a dead giveaway for this amazing fact: Shark skin is made
up of modified teeth. Yes, these skin plates share a common origin with teeth
and are basically miniature versions of them. Thus, shark skin is very tough but,

at the same time, the reticulated structure means it is also very flexible. The orientation of the scales, from nose to tail, means sharks are smooth and hydrodynamic in one direction and coarse in the other. If you are ever lucky enough to touch a shark, make sure you don't stroke backward or you may end up with a nasty case of shark burn—there is a silver lining: if that happens, you can claim to have been bitten by a shark, since the scales are miniature teeth!

Another mind-blowing feature that allows us to differentiate between sharks, skates and rays, and other aquatic creatures, is that the former group is able to sense extremely weak electromagnetic fields. Remarkable jelly-filled pores embedded in the skin on their head, and in especially high concentrations around the mouth and snout, can detect minute electrical currents and magnetic fields, with wide-ranging applications. From being able to sense the minuscule current generated by the heartbeat as their prey tries to hide under the sand to tracking the Earth's geomagnetic fields for navigation, this extra sense is a true superpower.

Kissing cousins

The class Chondrichthyes includes all sharks, skates, rays, and chimaeras. These animals are all very closely related (we will go into more detail in Chapter 3 about their extensive family tree), but here let's distinguish between skates and rays, chimaeras, and sharks.

The chimaeras, while once massively diverse, are now a relatively minor group of around 53, principally deep-water species that look nothing like their chondrichthyan cousins. They diverged from the shark and ray lineage some 400 MYA. Chimaeras have a tapered tail and a single gill flap on either side of their body that covers the gill slits.

The skates and rays, also known collectively as batoids since they are in the superorder Batoidea, subsequently diverged from the shark lineage about 230 MYA. The batoids then went their own way (evolutionarily, that is). This might sound like an awfully long time ago, but in the scope of evolutionary time, sharks and batoids are quite closely related. Batoids are effectively "pancake sharks," with a dorsoventrally flattened, or depressed, body and broad pectoral fins that have extended into flapping "wings" fused to the entire length of their body. Batoids still have five gill slits (or in the case of a single species, six), like most sharks, but they are located on their underside rather than on the flanks.

You might think it would be quite easy to differentiate a batoid from a shark, but the lines can become blurred, and there are many species of rays that look more like sharks and some sharks that resemble batoids. Angel sharks (family Squatinidae) and wobbegongs (family Orectolobidae), for example, could easily be mistaken for batoids, while the aptly named Shark Ray (*Rhina ancylostoma*), which is, in fact, a batoid, looks like an evolutionary stepping stone between the two groups. Many people struggle to tell the difference between the sawfishes (family Pristidae), which are batoids that resemble sharks, and the sawsharks (order Pristiophoriformes), which are true sharks. It is beyond the scope of this book to go into all the minute details of the differences between sharks and batoids, but a good rule of thumb is that if you are in an aquarium and happen to observe a fish from the side and you can see five or more gill slits on its flanks, it is a shark. If you cannot see the gills because they are on the underside of the body, it is a batoid.

OUR MAIN CHARACTERS

A biography must have a subject, but "sharks" are far too broad and varied a group to profile in one book. Instead, to tell the shark story, we will focus where appropriate on four species: the Great White Shark, Smallspotted Catshark, Sandbar Shark, and Spiny Dogfish (*Squalus acanthias*). While these four species will make repeated appearances, numerous other shark species will also have cameo roles, to provide you with the true breadth and wonder of the shark story. Our four species were selected not because they are particularly iconic (well, at least one is), but rather because they mostly represent the diversity of lifestyles, habitats, reproductive modes, and swimming styles of sharks as a whole. There is one exception here: we have not profiled a deep-sea shark, largely because they are, as scientists would say, data deficient; in other words, not enough is currently known to justify featuring one of them. Yet each of the four species also has its own unique story.

Our four featured sharks have been more extensively studied than many other sharks, so we know quite a lot about their lives. Yet this statement is misleading, since the obstacles to studying sharks (inadequate funding, technological impasses, the sheer size of some species, inaccessible habitats, and the fact that scientists are terrestrial air-breathers and sharks are not) translate into more questions about their biology, ecology, and behavior than answers. Good news for those of you wanting a career in shark science or conservation, but less happy news for sharks.

The Great White Shark

How could we not start with the absolute icon of sharkdom: the Great White Shark (*Carcharodon carcharias*)? This beast is the star of monster movies and horror books, the shark of our dreams and nightmares. Speed, agility, and strength, the Great White Shark is an enormous powerhouse of predatory perfection. The species is distinguished (if you will excuse us deviating from scientific accuracy) by the reaction of anyone seeing this shark in its natural environment for

3 ft

1 m

FEMALE GREAT WHITE SHARK
The Great White Shark is the paragon of powerful, predatory prowess—
a swift-swimming beast of mind-boggling proportions, and a true apex predator.

the first time: "Holy Cow!"). An adult Great White Shark clearly announces what it is, without the need to provide a list of how its parts add up.

Once you've recovered your breath, you might want to know this shark's distinguishing features. Great White Sharks have a long, conical snout and stout body, with long gill slits and a powerful, crescent-shaped tail. Also note the presence of a keel, an expanded lateral extension of the body, on the caudal peduncle (the area just in front of the tail fin). The Great White Shark's closest cousins are other sharks in the family Lamnidae, including the Shortfin Mako, Longfin Mako (*Isurus paucus*), Porbeagle (*Lamna nasus*), and Salmon Shark (*Lamna ditropis*), none of which approach the dimensions of the Great White Shark.

Growing to about 20 ft (6 m) and reaching 4,200 lb (1,900 kg), the Great White Shark weighs as much as a rhinoceros, yet is only two-thirds the mass of a Blue Whale's tongue. In fact, Great White Sharks are only the third largest shark, after the Whale Shark and Basking Shark, but they are the largest non-plankton-eating predatory fish. In 1945, a specimen measuring 21 ft (6.4 m) and weighing 7,328 lb (3,324 kg) was caught off the coast of Cuba, and this individual was deemed the largest reliable record of the species. The Great White Shark is found in all oceans in both coastal and oceanic waters of 54–75°F (12–24°C) and undergoes incredible migrations, in some cases across entire oceans.

As predators, Great White Sharks have few equals in the marine realm, in large part because of their ability to retain heat, which translates into increased muscle power, sensory acumen, and faster growth. While the Great White Shark is no Megalodon in size (these could measure as much as 60 ft/18 m), the adults occupy a broadly similar ecological niche as Megalodon as an apex predator, feeding at the very top of their food chain, predominantly on large marine mammals like seals and whales. Great White Sharks are known to breach when attacking seals from beneath, hurling their body several meters into the air.

Great White Sharks are so called thanks to the startling whiteness of their belly. It's thought the common name arose because, back in the day, most Great White Sharks were seen only when dead and lying belly-up on boats or fishing docks. The "great" of the common name is an unnecessary affectation because, firstly, its greatness is patently obvious without including the word in the name

GREAT WHITE SHARK DISTRIBUTION MAP
Great White Sharks are found globally in tropical and temperate coastal and oceanic waters of 54–75°F (12–24°C).

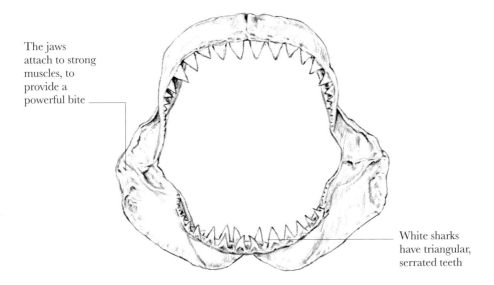

The jaws attach to strong muscles, to provide a powerful bite

White sharks have triangular, serrated teeth

THE REAL JAWS
The powerful jaws and triangular, serrated teeth of the White Shark have evolved to remove chunks of the flesh of their prey, including marine mammals, other sharks, and large bony fish.

and, secondly, the common name accepted by the scientific community is simply "White Shark." From this point on, we'll mostly use the accepted common name "White Shark."

This illustrates some major problems with the use of common names in general. White Sharks are also known as Great Whites, White Death, and White Pointers, depending on where you are in the world. To solve this confusing problem, scientists rely on only one binomial scientific name, which can be used no matter where you are from or what language you speak. For White Sharks, this is *Carcharodon carcharias*, from the ancient Greek words meaning "sharpened, jagged teeth" and "pointed." Together the two parts of the name describe the serrated, triangular shape of the White Shark's teeth. Recently, efforts have been made to assign unique common names to fish species. Although these will never replace the scientific names, they are useful. In order to make this book more enjoyable to read (especially for the nonscientists out there), we will include the most consistently used common names in the body of the text. For those interested in learning the formal nomenclature, there is a list of the scientific names of all species mentioned in each chapter at the back of the book.

In subsequent chapters, we will describe other aspects of the White Shark's biographical story, but despite being one of the most famous species, there are huge gaps in our understanding of the life history, physiology, and ecology of the White Shark. For example, shark biologists have not observed them mating or pupping, and have only recently learned about specific habitats vital to neo-nates (newborns) and juveniles—for example, off southern California, the west coast of the central Baja California Peninsula, in Mexico, and in the New York Bight on the United States East Coast. Contrary to public perception, the over-all population of White Sharks is not declining, although some regional popu-lations may be at risk. However, the future impacts of climate change and overall biodiversity loss do not give one cause for optimism.

The Smallspotted Catshark

Let's take it down to the opposite end of the spectrum, with the Smallspotted Catshark, also known as the Lesser Spotted Catshark or Sandy Dogfish (although it is not, taxonomically speaking, a dogfish at all—catsharks lay eggs, while dogfish give birth to live young). In fact, this shark has so many different

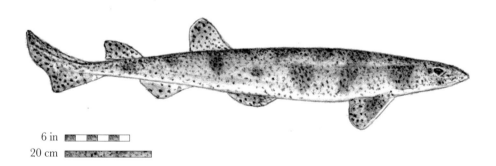

6 in
20 cm

THE SMALLSPOTTED CATSHARK
Every bit as much a predator as the White Shark, but with a menu
of much smaller prey that it can swallow whole, the Smallspotted
Catshark is a small, relatively abundant shark.

SMALLSPOTTED CATSHARK DISTRIBUTION MAP
*The Smallspotted Catshark is found over a wide depth range in its
Northeast Atlantic Ocean and Mediterranean Sea habitats.*

common names, in so many different languages, that it might be easier to list what it is not called. To scientists they are *Scyliorhinus canicula. Scyliorhinus* comes from the ancient Greek, meaning something like "small shark" or "sea dog." This was a somewhat derogatory term used to refer to many different species of small sharks in ancient times, much as some contemporary fishers call their unmarketable catch "trash fish." The Latin species name *canicula* is similar, literally meaning "little dog." There is disagreement regarding the rest of the name, with some experts thinking that the genus suffix comes from *rhinos* meaning "nose," while others consider it to come from *rhine*, meaning "rasp," in reference to the shark's jagged skin, which is so rough that it was once used as sandpaper or for grip on the hilts of swords.

The Smallspotted Catshark is an abundant species in the Northeast Atlantic and Mediterranean, preferring to live over sandy, gravelly, or muddy substrates, from shallow regions to depths of 2,624 ft (800 m). It is often found in both marine aquaria and fish markets in some areas (although it is not widely consumed across its range). As a result of this shark's broad regional distribution, relative abundance, small size, how easily it is collected as well as maintained in

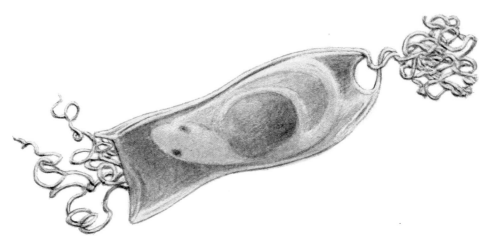

MERMAID'S PURSE
*Smallspotted Catsharks are among the 40 percent of sharks and rays that
are oviparous, or egg laying. Eggs are laid in pairs and will incubate for
as little as five months to as long as 11 months.*

captivity, and other biological attributes, the Smallspotted Catshark has been widely studied and could be considered a model species.

Although not fitting the archetype of a large, gray, swift-swimming, toothy shark, Smallspotted Catsharks are, in fact, members of one of the largest group of sharks alive today, the catsharks (family Scyliorhinidae). They reach a length of about 3.3 ft (1 m) and can weigh 3 lb (1.4 kg). Smallspotted Catsharks have a beautiful, marbled patterning across their yellow skin, with black and gray saddles, patterns that straddle their body, breaking up into spots with sparse white speckles. The long, slender body is topped by two dorsal fins set back far along the spine, near the tail. The upper and lower lobes of the caudal fin are less distinct compared to their faster-swimming relatives, with the upper lobe topped with a square, flaglike tip. If you are ever close enough, you can tell these little sharks apart from similar species thanks to the distinctive nasal furrows, grooves that reach all the way from the nostrils to the mouth, and which are found in several other species, such as Horn Sharks and Nurse Sharks. This anatomical adaptation enables the shark to smell or taste its chemical environment twice, first in the nostrils and then in the mouth.

Behaviors of the Smallspotted Catshark include curling up into a donut shape, and it was among the first sharks in which individual personality traits, including boldness (inquisitiveness) and shyness, were demonstrated. Females lay two eggs at a time and attach the curly extensions of the eggs to structures on the bottom of the seafloor, where they will incubate from six to nine months; the extremely slender hatchlings are quite fetching.

The Sandbar Shark

If we could select a single shark to profile as representative of sharks as a group, it might well be the Sandbar Shark (*Carcharhinus plumbeus*), even though it does not fit the description of a typical shark, which is less than 3.3 ft (1 m) long and lives in the deep sea. It does, however, represent what many people most commonly think of as a "shark," having a large, streamlined body, gray countershading into a whitish stomach, and a strong tail at one end and intimidating teeth at the other. Yes, the Sandbar Shark is a classic beast, with a gloriously high first dorsal fin and elegant swimming style that confers a sort of majesty on the species. The *Carcharhinus* in the binomial name comes from the ancient Greek, meaning "of or

10.5 in

30 cm

THE SANDBAR SHARK
Widely acknowledged as one of the most graceful of sharks, the Sandbar Shark is recovering from historic overfishing along the US Atlantic coast.

SANDBAR SHARK DISTRIBUTION MAP
The Sandbar Shark can be found worldwide in temperate coastal regions,
swimming above the bottom in waters shallower than 330 ft (100 m).

pertaining to a shark." However, *plumbeus* is Latin, directly translating to "made of lead" in reference to this shark's dark gray, sometimes almost bluish, coloration. The symbol for lead, Pb, comes from the same Latin root.

Sandies, as they are commonly called, are distributed circumglobally and also widely exhibited in large aquaria. The species has a compelling story of near-tragic overfishing and phoenixlike arising from the ashes in some regions, which we will discuss in Chapter 9. We call it the ideal training shark at Coastal Carolina University, in South Carolina, where author Dan Abel teaches, since it dominates our catch on longlines, is cooperative as our students collect data from it (although we have experienced minor bites on numerous occasions, due to having too cavalier an attitude), and survives beautifully after we tag and release it.

You can tell a Sandbar from other classic-looking gray sharks thanks to their prominent dorsal fin, which is noticeably larger than many closely related reef sharks, and their relatively short snout. They have saw-shaped upper teeth, perfect for grabbing and shearing their fish and invertebrate prey (crustaceans, squid, and so on). Fishery scientists also note that they have an especially prominent interdorsal ridge—a visible elevated line of tissue that runs between the first and second dorsal fins, along the shark's back. Male Sandbar Sharks commonly reach a length of 6.5 ft (2 m) from nose to tail, but the females grow larger than the males (around 28 in/70 cm bigger). As this species bears live young, the female's body needs extra space to carry her litter until birth.

Sandbar Sharks are one of the larger coastal shark species, living in shallow waters in estuaries and bays, often over muddy bottoms littered with (you guessed it) sandbars. Yet in most ecosystems they are not an apex predator and can become prey to bigger sharks throughout their range in temperate and tropical oceans across the world. Young Sandbar Sharks live in an area separate from the adults (known as a nursery habitat) and the mature sharks live in same-sex aggregations. This size and sexual segregation is common among many different species of sharks—there's more on that to come in Chapter 5.

Sandbar Sharks take about 15 years to reach maturity. They produce about eight pups every two years, a conservative reproductive strategy that hinders their ability to recover quickly from historical overexploitation. As a result of this overfishing, they are now categorized as Endangered on the International Union for Conservation of Nature (IUCN) Red List. Despite aggressive management of the targeted longline fishery for sharks along the United States East Coast, both Sandbar Sharks and their cousins the Dusky Sharks (*Carcharhinus obscurus*) have failed to regain their numbers fully, although the trajectory of their recovery is trending in the right direction.

The Spiny Dogfish

Another little critter which breaks the shark mold is the aptly named Spiny Dogfish, so called for the sharp spines that project from its two dorsal fins. Adapted from modified dermal denticles, the spines are a defensive weapon to protect these dogfish from predators. If caught, a Spiny Dogfish will arch its back to impale its captor, injecting a mild venom. Since it reaches only a little over 3.3 ft (1 m), there are certainly many larger animals, including bigger sharks, who might think twice about taking a Spiny Dogfish for lunch. The

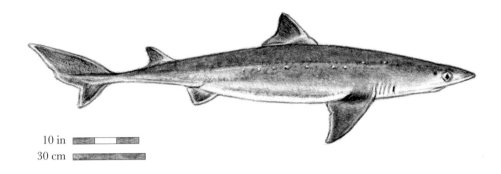

10 in
30 cm

THE SPINY DOGFISH
Named for the mildly toxic spines on its dorsal fins, this species has very
conservative life history characteristics and, despite an enormous population
size, has been overfished in parts of its range.

scientific name, *Squalus acanthias*, comes from two different ancient languages: *acanthias*, the Greek for "thorn" or "spine," and the Latin *squalus*, which was used to describe any species of shark in ancient times.

Spiny Dogfish are found in cold temperate waters of the North Atlantic and Pacific Oceans intertidally to depths of about 3,000 ft (900 m). They undergo seasonal migrations and can even enter freshwater areas for short periods. In fact, the species recently featured prominently in the press in the United Kingdom and even globally when specimens were caught in the River Thames in 2021, with the media sensationalizing the finding by focusing on the species' venomous spines and implying there was a public safety threat. While penetration with these spines into your hand can be excruciating, such injuries are not life-threatening. Spiny Dogfish school in huge groups, thousands strong, and they are even known to hunt in packs.

The Spiny Dogfish has a long, slender body, usually dark gray or brownish in color, sometimes scattered with small, white spots. They have no anal fin and a somewhat flattened head with large eyes. Their teeth are very similarly shaped in both the upper and lower jaws and, as the spiked peaks lean backward in the mouth, they come together to form a nearly continuous cutting edge—quite a formidable adaptation for feeding. Male and female Spiny Dogfish reach noticeably different maximum sizes, with the females significantly larger than the males, 42 in (107 cm) and 35 in (100 cm), respectively. This sexual dimorphism occurs because the mother carries her young for such a long period of time when pregnant. We'll go into more detail on this and other types of sexual dimorphism in Chapter 7. The species is the poster child for slow, or conservative, life history characteristics. In addition to its long gestation period, it takes Spiny Dogfish as long as 20 years to mature, after which they may give birth to only four to six pups every other year.

The Spiny Dogfish is recovering from being overfished in the Northwest Atlantic in the recent past and is also listed as Endangered in the Mediterranean Sea and throughout Europe on the IUCN's Red List. In the 1930s, Spiny Dogfish were used as a source of vitamin A, and if you have ever eaten fish and chips in the United Kingdom, you have likely consumed this species, which is known there as Rock Salmon. The Northwest Atlantic stock of the Spiny Dogfish plummeted before fishing was restricted and after a period of rebuilding, mature females and pups may be declining again as a result of overfishing.

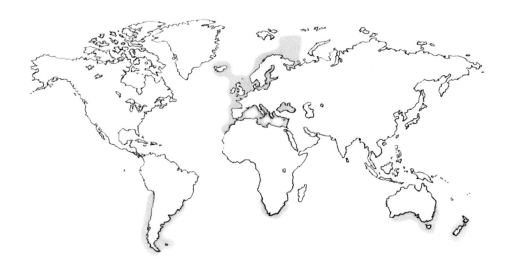

SPINY DOGFISH DISTRIBUTION MAP
*The Spiny Dogfish is found in the intertidal zone to as deep as 900 m
(2,950 ft) along coastlines. It is endangered in the Mediterranean Sea and
northern Europe, but remains plentiful in other areas.*

Now that you have learned a little more about sharks, their current biodiversity, and our four fantastic featured species, if you were asked again where you would like to be transported for a sharky encounter, would your answer still be the same? If you have not changed your mind, maybe taking a trip back in time—some 450 million years ago, to when the earliest ancient shark lineages first arose—could sway your answer. We'll open the window to that era in the evolutionary history of sharks in the next chapter. Pack your bags!

2

ROOTS

WHERE THE STORY BEGINS

More than 400 MYA, when the climate was noticeably warmer than it is now, before the supercontinent Gondwana had completely formed, and North America, Greenland, and Europe were all conjoined as the supercontinent Laurasia, 85 percent of the Earth was covered in water. Millions of years before the first dinosaurs arose, even before tetrapod animals (represented now by amphibians, reptiles, birds, and mammals) had crawled out of the water, and before trees had evolved on land, the oceans teemed with life. Filling every imaginable niche from shallow, nearshore waters to the abyssal ocean plains, armored fishes ruled, trilobites and gastropods flourished, coiled ammonites cruised the depths, and bivalves and burrowing organisms lived amid the sponges, corals, and stromatolites (layered deposits of limestone formed by microorganisms).

Often dubbed the Age of Fishes, this period, known as the Devonian (some 420–360 MYA, or perhaps 20 million years or so earlier), was probably when the first early ancestors of sharks appeared in the oceans. While it might sound as if this is where a shark biography should begin, if we started here, you would miss the beginning of the story because sharks did not appear out of thin air (or thin water). Although evolution is sometimes sudden, it is mostly subtle and acts slowly. So, it is only when we look back over millions of years that we start to see the major events and vital patterns in how modern animals came to be. To really understand how sharks arose, we need to go back much earlier— before sharks were even a twinkle in evolution's eye.

Life with more bite

Evolution has given rise to a myriad of different life-forms on our planet: everything from corals to dinosaurs, from flowering plants to elephants, emus, whales, bacteria, and giraffes. All this life began with a single-celled microorganism that lived in our oceans around 3.7 billion years ago. These tiny heroes subsequently adapted and evolved into different species of microbes and

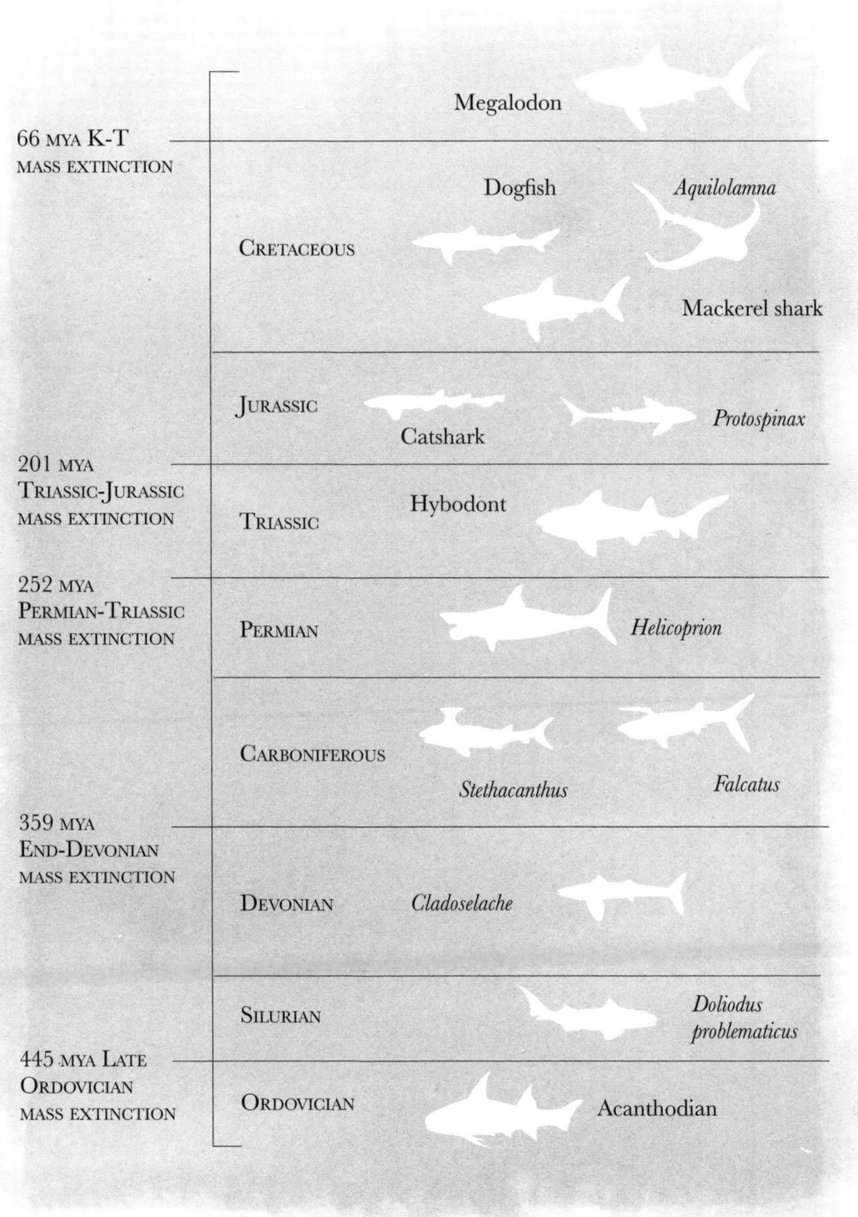

	Megalodon	

66 MYA K-T
MASS EXTINCTION

CRETACEOUS

Dogfish · Aquilolamna

Mackerel shark

JURASSIC · Catshark · Protospinax

201 MYA
TRIASSIC-JURASSIC
MASS EXTINCTION

TRIASSIC · Hybodont

252 MYA
PERMIAN-TRIASSIC
MASS EXTINCTION

PERMIAN · Helicoprion

CARBONIFEROUS

Stethacanthus · Falcatus

359 MYA
END-DEVONIAN
MASS EXTINCTION

DEVONIAN · Cladoselache

SILURIAN · Doliodus problematicus

445 MYA LATE
ORDOVICIAN
MASS EXTINCTION

ORDOVICIAN · Acanthodian

SHARKS THROUGH TIME

*Sharks' ancestors have been around since the Ordovician Period, and their
lineage has survived through several mass extinction events.*

eventually gave rise to plantlike cyanobacteria, which were able to photosynthesize and, therefore, to generate the oxygen in our atmosphere. In due course, life advanced to be multicellular and became more complex. Then animals appeared: sponges arose, followed by worms, jellyfish, and eventually early fish around 530 MYA.

Despite their modern-day counterparts coming in many different shapes and sizes, all sharks and their relatives actually derived from one single common ancestor. The lineage that gave rise to our contemporary sharks first appeared some 450 MYA or so. At this point, the cartilaginous fishes, including the sharks, skates, rays, and chimaeras (if you remember, these are known collectively as chondrichthyans) branched away from the bony fishes (osteichthyans).

A commonly cited figure is that the first sharks arose 440 MYA. To some extent this is true, as the earliest examples of sharklike dermal denticles can be dated to that time, but this assertion is a little misleading. It might be a stretch

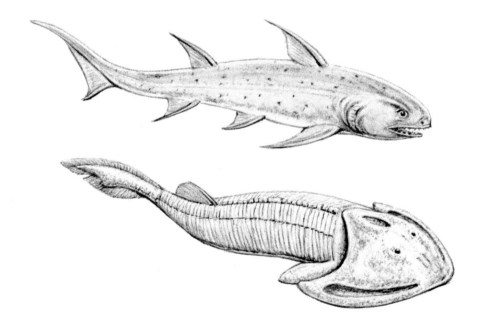

ACANTHODIAN AND JAWLESS FISH
A typical acanthodian (top) compared to a jawless fish of the time.

to say that fish living during this period were genuine sharks, since the question of which sharklike creature had enough of the required characteristics to make the claim that it was undeniably the earliest shark remains unanswered.

What is not unresolved is an event that occurred during the end of the Ordovician Period, no later than 450 MYA: the evolution of jaws. Jaws were a game-changer, shifting the path of vertebrate evolution forever. The two earliest groups of jawed fishes, placoderms and acanthodians, were no longer limited to vacuuming up their dinner from the ocean floor like their jawless ancestors; they were now able to bite, chew, and crush their food, and manipulate prey in their mouths. The presence of jaws changed their entire approach to life and expanded their habitats, opening up a wide range of different niches which could now be explored and exploited. This led to an explosion of new species that defined the Age of Fishes. In fact, every major clade (taxonomic group) of fishes—sharks included—that we know and love today arose during this magical time.

THE ONE AND ONLY

It is difficult to say with any certainty when exactly the first shark appeared for several reasons. First, after death, cartilaginous skeletons of chondrichthyan fish degrade quickly and only heavily mineralized parts like the vertebrae fossilize. This means only patchy examples of ancient sharks exist in the fossil record. Complete shark skeletons are very rarely discovered, leaving paleontologists to rely almost solely on teeth and skin scales, which also fossilize relatively well thanks to their hardened coatings.

What's more, it is difficult to categorize early shark ancestors because evolution is mostly a gradual process, with subtle adaptations slowly shifting into larger changes over many millions of years. In more derived species (those that arose later on the evolutionary tree), it is easy to tell a chondrichthyan from an osteichthyan because skeletons of the former are made up entirely of hardened cartilage, which never ossifies into bone. However, during the early stages, when these branches first separated, there were many crossovers and shared features between the groups, which were challenging to untangle.

One such group was the acanthodians, which were prematurely dubbed the "spiny sharks," thanks to their sharklike shape and multiple pairs of fin spines. Their resemblance to sharks—with streamlined bodies, skeletons made of cartilage, and functional jaws filled with sharp teeth—masked the fact that acanthodians were actually a mosaic of both chondrichthyan and osteichthyan traits. No other shark, past or present, possesses multiple pairs of spined fins. However, two recent scientific papers concluded that acanthodians are indeed the true stem chondrichthyans. The other group, which at one time was thought to have given rise to sharks (and bony fishes as well), was the placoderms, a highly successful assemblage of dominant predators with a bony endoskeleton and a hinged, ossified, helmeted head. Placoderms, however, exhibited some anatomical dissimilarities to chondrichthyans and bony fishes that disqualified them from being considered ancestral to both groups.

After the rise of the acanthodians during the Ordovician Period (485–444 MYA), the fossil record reveals that there were many different species with sharklike forms, but it is hard to state with absolute certainty which—if any—was actually the first true shark. Most notably, it is difficult to determine whether some of the iconic early sharklike creatures were really sharks or part of a closely related sister-group, the chimaeras. Rather ominously dubbed ghost sharks, rat fish, or spook fish, but formally known as holocephalans, chimaeras branched away from the rest of the chondrichthyan fishes around 421 MYA, but there were many species along the way which were a bit like sharks and also a little like chimaeras.

We may never be able to unequivocally state which was the first true shark, and it is probably more realistic to identify a collective or a continuum of species which eventually evolved into sharks as we now know them. Scientists refer to these animals as stem chondrichthyans because they were neither true sharks nor genuine chimaeras, but have some features of each.

EARLY SHARK EVADING *DUNKLEOSTEUS*
During the Devonian Period, nimble sharks known as Cladoselache *were
adapted to evade heavy, armored predators called* Dunkleosteus.

One of the oldest sharklike stem chondrichthyans is the appropriately named *Doliodus problematicus,* which lived some 400 MYA. These creatures were a problematic mish-mash, with a sharklike head (including teeth, braincase, and jaws), but the body of an acanthodian (with several pairs of fin spines). Paleontologists suspect this creature might be the missing link between the ancient acanthodians and sharks proper.

Slightly later—around 390 MYA during the Devonian Period—a group known as the cladodont sharks arose. Within this group, members of the genus *Cladoselache* are commonly referred to as the first real sharks, but others argue they were actually more likely chimaeras. This group of creatures was unusual because their skin was thin and relatively fragile compared to the hardened scales which make other sharks' skin so tough. But this was not a weakness! On the contrary, it allowed *Cladoselache* to be nimble and quick. At the time, the mighty, armored placoderm fish known as *Dunkleosteus* roamed the oceans and the light frames of *Cladoselache* probably allowed them to evade these larger, heavier predators.

A GOLDEN AGE

While the process of evolution is awe-inspiring, it can also be incredibly violent, continuously weeding out those too weak to survive. Even those organisms that pass the test of survival of the fittest often fall prey to massive extinction events periodically devastating the planet's biota. Near the end of the Devonian Period, a cascade of several mass extinction events over some 500,000 to 25 million years changed the Earth forever.

Whether this parade of catastrophes was caused by a destructive meteor, a major crash in ocean oxygen levels, sea level shifts, climate change, or some fiery supervolcano, we may never know for certain, but we do know that the marine world was devastated. During the End-Devonian Extinction (also called the Kellwasser Event) 371–359 MYA, about 70 percent of all marine life went extinct. Reefs almost completely disappeared and whole groups of trilobites (extinct marine arthropods), shellfish known as brachiopods, and the placoderms were completely wiped out. But all this death actually made way for the sharks to rise.

Sharks rule

What followed was the Carboniferous Period (some 359–299 MYA). At this time, tectonic drift had formed two major oceans, where warm, shallow waters often flooded over the new supercontinent, Pangaea. The climate was continuously hot and humid, with indistinct seasons. On land, vast bogs and dense swamp forests stretched across Pangaea, while high atmospheric oxygen levels fueled unparalleled growth of insects to gargantuan sizes—millipedes the length of cars and dragonflies with wings measuring 3.3 ft (1 m) across. In the oceans, sharks ruled.

Bony fishes and the early sharks had not only survived, they had thrived! With the extinction of many of the primitive jawless fishes and the armored placoderms, sharks and their relatives were freed from a huge competitive burden and had more space to flourish. Amid recovering reefs hosting starfish, urchins, marine worms, and sea snails, chondrichthyan fishes exploded into a multitude of different (often fantastically bizarre) forms and ecological niches. In fact, this period is known as the Golden Age of Sharks.

At the beginning of this period, some 340 MYA, there lived a group of fishes of the genus *Saivodus*. Reaching incredible sizes, up to 26 ft (8 m) from snout to tail, and boasting the classic streamlined, multi-finned shape so recognizable as shark-like today, these creatures can undeniably claim to be sharks (maybe even *the* first true sharks). Yet, later in the Carboniferous, there was also a multitude of other fantastical sharklike animals that may have been true sharks or perhaps more accurately chimaeras. One such group was the genus *Stethacanthus*. Arising around 330 MYA, male stethacanthids boasted a truly fantastical, frilled anvil resembling a wire scrub brush that stuck straight out from the top of their heads. This adapted dorsal fin, which scientists muse may have been inflatable so it could become engorged during courtship displays, is thought to have been a sexual ornament that arose to attract females. No modern sharks possess anything like this.

Similarly, a little later on in the Carboniferous (some 320 MYA), male *Falcatus* sported a unicorn-like, elongated first dorsal spine that stuck out from the front of their face. Like the anvil of *Stethacanthus*, the spines of *Falcatus* were

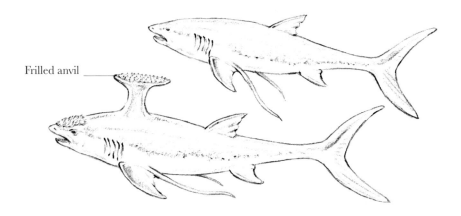

Frilled anvil

SEXUAL DIMORPHISM OF *STETHACANTHUS*
The frilled anvil of the male Stethacanthus *(lower)*
likely evolved to attract a female for mating.

sexually dimorphic (different in males and females). Scientists think the spines evolved as a display of strength that *Falcatus* used to attract a mate and over time sexual selection favored larger spines, causing the appendage to become ever-more pronounced.

Give it a whorl

Around 299 MYA, the oxygen-rich Carboniferous gave way to the Permian Period. After a massive decline in atmospheric carbon dioxide led to pronounced climate change, drying, and major glaciation, rainforest ecosystems collapsed across the supercontinent and life on land changed forever. In the single super-ocean that spanned the entire globe, encircling Pangaea, waters teamed with snails, spiral-shelled mollusks known as ammonoids, and graceful, wafting feather stars called crinoids.

It is probably during this period that the most iconic of the sharklike ancestors lived (although it may more accurately be described as a chimaera): the magnifi-

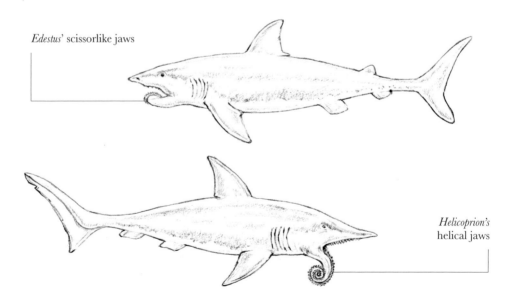

Edestus' scissorlike jaws

Helicoprion's helical jaws

JAWS OF PERMIAN SHARKS
During the Permian Period, the extinct sharklike fishes
Helicoprion *and* Edestus *had crazy jaw morphologies.*

cent group in the genus *Helicoprion*. Arising around 290 MYA, these bizarre fishes sported a remarkable whorled jaw that projected as a tooth-lined spiral from the front of their face. This feature, which gave the shark a unique, chopping bite, arose because these animals never lost any of their teeth, as modern sharks do. Yet new ones didn't stop growing either. This resulted in the newly growing teeth at the back of the jaw pushing the older teeth out to the front, forming a wonderfully weird helix. This so baffled paleontologists when they first discovered it that they wondered if it was some heretofore undiscovered ammonite.

During this time there also existed a group of close relatives, known as *Edestus*, or more commonly the Scissor-toothed Shark. These creatures also never lost teeth, but in their case, this caused the jaw to extend and project forward and outward. Scientists think this made *Edestus* perfectly adapted to hunting because, unlike any other known shark, the remarkable scissor-mouth meant they could move their body up and down to slash prey with staggering force.

A NEW HOPE

Around 252 MYA, unimaginable volcanic eruptions caused catastrophic global warming and massive changes to ocean chemistry, which resulted in the largest extinction event to date. During what has become known as The Great Dying (also referred to as the Permo–Triassic Extinction Event, or the End-Permian Extinction), as many as 81–96 percent of all marine species, including corals and trilobites, were completely wiped out and 70 percent of all land animals were also killed off. This marked the end of the remarkable diversity enjoyed during the Golden Age of Sharks.

Edestus and the cladodont sharks did not make it, yet several early shark lineages did survive this catastrophic extinction event. Experts suspect that they retreated into deep-sea, offshore refuge areas to ride out the changes going on around them. It is during this period (about 200 MYA) that we start to see evidence of the first

XENACANTHS AND HYBODONTS IN RIVER ECOSYSTEMS
Ancient sharks thrived in freshwater river ecosystems during the Triassic Period.

modern sharks, known as neoselachians, in the fossil record. Alongside them, their sister groups, the hybodont sharks and the xenacanthiforms, also survived into the new post-extinction period, flourishing in the brackish (somewhat salty) and freshwater environments of rivers, as well as in the oceans.

Hybodonts had the familiar sharklike form, yet they were incredibly diverse, especially in their teeth, so much so that many dental fossils from this time remain unidentified or unnamed. Hybodonts were the dominant group during the Triassic and into the Early Jurassic Period (240–230 MYA). However, the hybodonts eventually began to be overtaken by the neoselachians toward the end of the Jurassic.

Sadly, the end of the Triassic (around 228–201 MYA) saw another global extinction event called the Triassic–Jurassic Extinction. A gradual change in climatic conditions, changing sea levels, and ocean acidification over approximately a 30-million-year period yet again caused the extinction of up to 75 percent of all marine and terrestrial animals.

Jurassic shark

But with every end there is a new beginning and after this tragedy, we see the rise of some of the most iconic animals in the history of our Earth: the dinosaurs. Is there anyone on our planet who doesn't know about this period in geohistory? Even its name—the Jurassic Period (201–145 MYA)—is familiar thanks to the blockbuster movie franchise. Yet many people don't realize that this era boasted not only gigantic lizards, but also magnificent sharks.

While many shark lineages were once again wiped out in the preceding mass extinction, some persisted and enjoyed quite a long period of relative stability afterward. Sharks subsequently became very common in the oceans, and they began to diversify readily and rapidly into some of the forms familiar to us today. In fact, the majority of modern sharks and rays originated during the Early Jurassic Period, with at least five of today's nine major groups (taxonomically known as orders) all arising during this time.

For example, around 195 MYA, the first hexanchid sharks appeared. This lineage gave rise to the modern hexanchiform sharks, including the six-gills and seven-gills that are still swimming in our oceans today. In fact, some of the

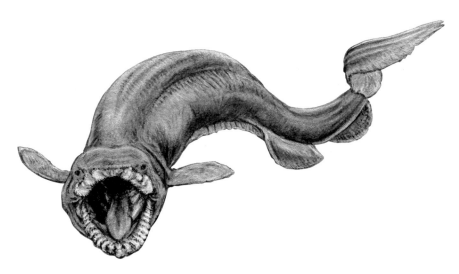

FRILLED SHARK
Frilled Sharks are known as "living fossils," as they have been swimming in the oceans since the time of the dinosaurs.

49

species in this group, such as the Frilled Shark (*Chlamydoselachus anguineus*), have been around for so long that they are known as "living fossils" and have remained basically unchanged for millions of years.

Around 168 MYA, the first catsharks appeared, ancestors of today's Small-spotted Catshark (*Scyliorhinus canicula*) and their relatives. At a similar time, early White Shark (*Carcharodon carcharias*) ancestors known as *Paleocarcharias* also came into being. The earliest ancestors of the Spiny Dogfish (*Squalus acanthias*) first appeared some 125 MYA. Then the first mackerel sharks (order Lamniformes) and stingrays appeared, some 135–130 MYA.

During this period, sharks developed more flexible, protruding jaws, which meant they could handle larger prey, even animals bigger than themselves. The speed with which they could swim also increased and, by 100 MYA, the so-familiar, large-bodied, speedy, modern sharks had come into being.

THE EAGLE SHARK
Aquilolamna *evolved a soaring style of locomotion millions of years before the batoids developed the same swimming adaptation.*

It was around 185 MYA that the earliest types of rays, known as *Protospinax,* first appeared. Scientists recently made an extraordinary discovery regarding the evolution of early rays, after finding a fossil of a new species of shark called *Aquilolamna* (*Aquilolamna milarcae*). Living alongside the dinosaurs around 93 MYA, the aptly named Eagle Shark had a broad mouth (similar to many rays), which scientists think was used for filter-feeding. It had a large tail fin (much like a modern shark), which it used to swim, but also sported elongated, winglike pectoral fins, spanning a whopping 6.3 ft (1.9 m) across. Thus, it would seem that these sharks "flew" through the water, using their fins for propulsion and maneuverability, much like modern-day manta rays.

This finding is remarkable because it changes everything we thought we knew about when this mode of swimming first appeared and how this adaptation arose. *Aquilolamna* proves that soaring arose as many as 30 million years before the ancestors of today's manta and devil rays (family Mobulidae) popped up with the same body plan. As this shark is thought to be more closely related to modern Lamniformes (early relatives of the White Shark) as opposed to rays, this means that "wings" for underwater flight did not evolve just once but multiple times over the course of evolutionary history.

Scientists suspect that *Aquilolamna* was wiped out by the K-T Extinction Event (aka the Cretaceous–Tertiary Extinction) that occurred 66 MYA. While you might not have heard the name, this extinction event is familiar to us all because it signaled the end of the dinosaurs and heralded the ascendancy of mammals. When a mighty asteroid struck the Earth, creating an impact crater over 90 miles (150 km) wide, enormous amounts of dust and ash were flung into the atmosphere, causing an immediate global climate shift. This killed 75 percent of all species, including many plants, all ammonites, mosasaurs (a group of extinct aquatic reptiles), and, of course, the dinosaurs. In the oceans the vast majority of very large predatory sharks and huge numbers of rays were driven to extinction, with about 85 percent of all species dying out. The diversity of larger-bodied members of the order Lamniformes was especially reduced.

Off the scale

This brings the total of mass extinctions that the sharks have managed to survive to no less than four. What's more, recent scientific research has shown that another catastrophic event may have specifically affected sharks. After developing a new method for analyzing fragments of teeth and fossilized skin scales known as ichthyoliths within seafloor sediments, scientists have discovered a heretofore unknown shark extinction that happened as recently as 19 MYA. We now know that during the Miocene Epoch (23–5 MYA), shark abundance dropped by as much as 90 percent and there was a 70 percent reduction in shark diversity across many different lineages all around the globe. Especially impacted were pelagic species of sharks—those that live in offshore habitats, far from land.

It seems that these families never fully recovered from such catastrophic declines. While the modern shark lineages began to resurge some 2–5 million years after this mass extinction, the loss of so many different species of sharks completely shifted the makeup of shark communities and this has reverberated through evolutionary lineages all the way through to the present day. As a result, our modern sharks, as fascinating and diverse as they are, may actually represent only a tiny fragment of what they once were.

The Meg

We couldn't possibly write a biography of sharks, telling the story about ancient shark ancestors, without devoting some time to the magnificent, mighty Megalodon, a beast which lived from 23 to about 3.2 MYA. Known to scientists as *Otodus megalodon*, the name literally means "mighty teeth." Megalodon was the largest macropredatory shark that ever lived. In fact, it was *the* largest predatory fish that ever lived. A single tooth alone could reach 7 in (18 cm) in length and their jaws could open wide enough to engulf the front of the average car!

As with other chondrichthyans, what we know about Megalodon has predominantly been learned by studying their fossilized teeth. Through a method known as allometry, scientists can look at the ratios of body parts of shark species that are alive today and use this to calculate how big an extinct shark would have been, based only on measurements from a tooth. However, as this

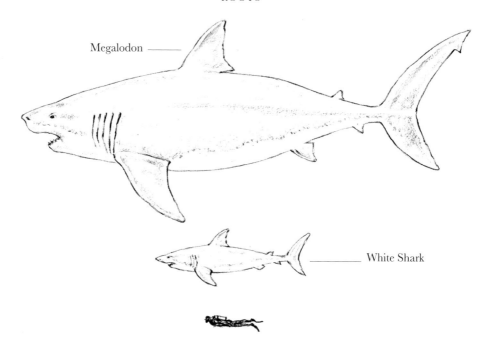

Megalodon

White Shark

MEGALODON COMPARED TO THE WHITE SHARK
The extinct Megalodon could reach sizes of 60 ft (18 m) in total length,
dwarfing even the largest modern-day White Sharks.

is not an exact science, we are not completely certain how large a size the Meg could achieve. As the most recently accepted length is about 60 ft (18 m), Megalodon was truly gargantuan. Imagine a shark the length of a semitrailer swimming past you (but not toward you) in the water.

Megalodon evolved to become so much larger than its ancestors, thanks to a size-based arms race between predator and prey. Over millions of years, as prey evolved to be larger and thus more difficult to hunt, predators evolved to become bigger ... prey then became larger still ... and so on. This vicious cycle pumped body forms to enormous sizes over the span of millions of years, culminating in the incredible Megalodon. This gigantism may, in fact, have contributed to Megalodon's downfall. Toward the end of the Meg's reign, an ancient climate change event caused significant global cooling. As this meant a lot of the ocean was now too cold for them, this once global, cosmopolitan predator's habitats constricted hugely and many of their favorite prey shifted their distributions as well. Concurrently, large predatory whales and the White Shark arose, and began competing with Megalodons for any of the food supplies that remained. Their wider thermal tolerance and more modest caloric requirements (thanks to their smaller size) meant that White Sharks were able to outcompete the Megalodon for prey, so eventually, they prevailed, and their lineage persisted, whereas Megalodon was driven to extinction.

Having arisen from a previous ancestor known as *Otodus obliquus* some 23 MYA, Megalodon was a part of the order Lamniformes, which today includes White Sharks, Basking Sharks, and Sand Tigers (more on this in chapter 3). At one point, scientists thought the Meg's closet next of kin was the White Shark, but we now know that this is not the case. In fact, studies of tooth morphology have elucidated that Megalodon was actually more closely related to modern-day Makos. This makes the Shortfin and Longfin Makos (*Isurus oxyrinchus* and *I. paucus*, respectively) Megalodon's closest living relatives and means Megalodon is not the ancestor of our contemporary White Sharks, as their lineages actually diverged around 100 MYA.

A question that shark scientists often get asked is, "Are Megalodons definitely extinct?" and the answer, unequivocally and without any uncertainty is "Yes"; Megalodons are definitely gone. Rigorous statistical analysis of the dating of Megalodon fossils tell us that they had died out no later than about 3.2 MYA. There have been no samples of Meg teeth or any other body part, fossilized or fresh, found anywhere in the world that are younger than this. If Megalodon was still around, we would certainly expect to find something, considering how prevalent they were in the fossil record in previous periods. What's more, we also never see any evidence that a gigantic macropredatory shark is living somewhere in our oceans, unseen; no huge bites on whales or other potential prey and no observations of the beast itself. All "sightings" of Megalodons have in fact been proven to be other species, like White Sharks, mako sharks (*Isurus* species), and often Basking Sharks (*Cetorhinus maximus*), which actually look nothing like the Meg. While it is true that our oceans are massive and largely unexplored, this does not mean Megalodons could still be out there. There is simply too little food in the deep-sea, offshore environments to support such an enormous predatory animal and they would not be able to survive the cold of the abyssal regions, like the Mariana Trench. Don't believe all the conspiracy theories you find on the internet or put too much stock in monster movies that are designed only to entertain, not inform; there is no doubt at all that Megalodon is, perhaps sadly, as dead as the dodo.

IT'S NOT OVER

Many people think of evolution as something that happened in the deep past, but on the contrary, it is a never-ending process. Both extinct ancient sharks and their extant counterparts are part of a continuous story of adaptation, disaster, change, and survival against all the odds. For example, the order Carcharhiniformes, to which Sandbar Sharks (*Carcharhinus plumbeus*) belong, first appeared in the Jurassic Period and modern mako sharks have been around for only some 60 million years. Although this might sound like a long time ago, it means these sharks and many others alive today are relatively new kids on the block compared to the chondrichthyan lineage as a whole. And ongoing environmental shifts, genetic mutations, and selection pressures continue to drive changes in shark morphology, habitat, and behavior to this day, although some of these changes (morphology, for example) operate on longer time scales than others (such as habitat changes).

New and improved

One example of evolution in sharks, a rather atypical one, is that of the Epaulette Sharks (*Hemiscyllium ocellatum*) in Australia, which have evolved to "walk" short distances on land. When the tide recedes, and small tide pools become isolated by stretches of sand and coral heads, these little sharks use their strong fins to move between submerged areas. Not only has their fin morphology allowed for this bizarre enhancement, but they have also evolved a remarkable hypoxia tolerance to do this—they are able to survive in very low oxygen.

While they are exposed to the air and unable to ventilate, Epaulette Sharks can widen their blood vessels to lower their blood pressure, which ensures oxygenated blood continues to reach the most important organs like the brain and heart. They have effectively evolved to be unparalleled breath-holders. Take as another example the hammerheads (family Sphyrnidae). This is one of the newest shark families, evolutionarily speaking, having arisen a mere 35 MYA. These sharks are absolutely iconic, boasting an otherworldly, elongated and flattened, mallet-shaped face known as a cephalofoil. These fantastical features may be cool to look at, but they also confer some adaptive advantage to the sharks which enhances their ability to survive—a remarkable feat of evolution. As the appendage is formed by the lateral expansion of cartilage housing the sensory regions, the cephalofoil provides the hammerheads with a marked advantage in sensory perception. Compared to other sharks, the elongated distance across their face gives them an enhanced sense of smell, better electrosensory perception, an enlarged visual field, and excellent binocular vision.

All these super senses mean that hammerheads are superbly evolved for their lifestyle. As if that wasn't enough, the strange head can also be used to pin prey like batoids to the seafloor while the shark consumes them. It can act as a forward rudder facilitating sharp turns. However, this incredible specialization might doom these remarkable sharks eventually. Scientists are now concerned that hammerheads are so specialized that they have been backed into a corner, evolutionarily speaking, and in today's anthropogenically driven, changing world, they may not be able to continue to adapt fast enough to their shifting environment.

* * *

With sharks continuing to evolve subtly as you read this, some species will probably very sadly be driven to extinction, but who knows what sharks might look like and what they could be doing a hundred million years from now?

3

THE SHARK
FAMILY TREE

WE ARE FAMILY

To construct a family tree, one must know the family members. So, we start this chapter on the shark family tree with a conundrum: shark taxonomists are unsure how many species of sharks even exist. In other words, they do not know all the family members. Today scientists have described no fewer than about 1,250 elasmobranch fishes (the sharks and batoids), but new species are being added every year. It is thus challenging to put a number on how many different types of sharks there really are.

In nature, a species consists of individuals that reproduce only with each other to produce viable and fertile offspring—that is, progeny that are basically healthy and capable of breeding when they mature. Each of the four sharks featured throughout this book is a distinct species. This means that, while a Great Dane is theoretically capable of reproducing with, say, a Chihuahua, since they are (perhaps incredibly) both in the same species (*Canis familiaris*), none of our shark species could do the same with each other.

You'd think it would be a relatively easy to work out the number of shark species, since distinct species don't look alike, right? That may be true for some organisms, but two different organisms can look identical and constitute different species. Consider the Carolina Hammerhead (*Sphyrna gilberti*), a sister species to the Scalloped Hammerhead (*S. lewini*), with whose range it overlaps along the coast of South Carolina in the southeast United States. The physical appearances of the cryptic species, the Carolina Hammerhead, and the established, named species, the Scalloped Hammerhead, are indistinguishable. The main anatomical difference between the two is that the Carolina Hammerhead has fewer vertebrae, but you'd need an X-ray machine or other imaging device to detect this. Yet, in 2006, a team of researchers from the University of South Carolina discovered that what shark biologists had aways called Scalloped Hammerheads were actually two genetically different species.

How then do taxonomists discriminate between species? Recent developments in the application of modern molecular biology have led to discoveries like the Carolina Hammerhead–Scalloped Hammerhead distinction, and these discoveries are occurring so quickly that it is making the heads of taxonomists spin as they attempt to confirm whether what molecular biology unearths in the lab also applies in nature.

Another explanation for the uncertainty around the number of shark species is the lag in assessments of underexplored areas, most notably the deep sea, but also of remote locations in sparsely populated areas. Therefore, when a new species of shark is discovered, this doesn't mean it spontaneously arose right under our noses. Rather, it has existed for some time, but it has simply evaded detection by researchers.

Where have you been all my life?

New species are discovered daily over a wide array of taxa across the spectrum of life, from amoebae to hedgehogs, so it isn't really surprising that new species of sharks are also being revealed. In the last ten years, around 54 new shark species have been identified, plus 100 batoids (rays and skates) and seven chimaeras. Some newly discovered sharks, besides the Carolina Hammerhead, include Lea's Angel Shark (*Squatina leae*), Halmahera Epaulette Shark (*Hemiscyllium halmahera*), Painted Hornshark (*Heterodontus marshallae*), Ridged-Egg Catshark (*Apristurus ovicorrugatus*), and two species of sawshark (not to be confused with sawfish, which are batoids), Kaja's Sixgill Sawshark (*Pliotrema kajae*), and Anna's Sixgill Sawshark (*P. annae*). Stunning beasts all, to be sure, but no large, legendary counterparts to the White Sharks (*Carcharodon carcharias*) or Tiger Sharks (*Galeocerdo cuvier*) have been added to the list recently.

What is the current best guess for the number of shark species? In 1984, the authoritative book on sharks, Leonard Compagno's *Sharks of the World*, listed 342 species of sharks. As we write, estimates from reliable and authoritative sources pin the number of taxonomically valid shark species at between 505 and 603, with batoids at 643–822 and chimaeras at 50–57, but by the time this book goes to press, it is likely that half a dozen new species will have been discovered and the numbers will already be out of date. As taxonomists validate these putative new species, we will continue to adjust our numbers.

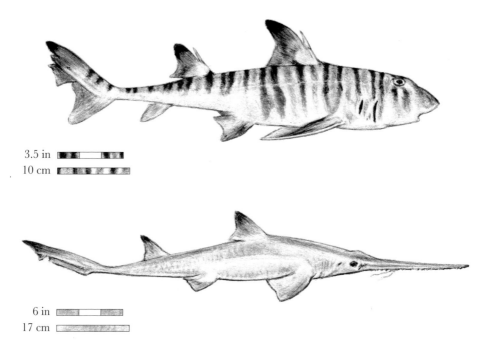

3.5 in
10 cm

6 in
17 cm

RECENTLY DISCOVERED
*The Painted Hornshark (top) and Kaja's Sixgill Sawshark, from
NW Australia and Madagascar respectively, are two species only
recently known to science.*

You may be surprised to hear there are so many species of sharks, but is this actually a high number? For insight into that question, look no further than the bony fishes, whose number of species may exceed 35,000. Among these bony fishes, there are more members of the carp/minnow family (around 3,000 species) than all the sharks, batoids, and chimaeras combined. So, while yes, there is prodigious diversity of sharks in terms of size, shape, behaviors, adaptations, habitats, and so on … no, their biodiversity actually pales in comparison to many other vertebrate groups. Biodiversity, however, is not the same as success (there is but a single living species of humans, although given the way we treat each other and the planet, maybe the word "success" doesn't apply to us), and the success of sharks as a group, and their outsized ecological importance, cannot be denied.

The shark diaspora?

You are likely wondering why there are relatively few species of sharks. Species form in many ways, but one of the main modes involves separating part of the population from an existing species, after which the newly separated group may diverge from the original population and given enough time—*voila!*—become a new species. The answer to why there are so few species of sharks plunges us to the depths of what it is to be a shark; that is, the shark's approach to life.

As a group, sharks and their close relatives have conservative life history characteristics, including slow growth, late maturation, long gestation periods, and relatively small numbers of young. All these features work against making new species. For instance, a key way for the separation of a population to occur is through mechanisms that facilitate dispersal. Bony fishes do a far better job of this than sharks. Some bony fish lay many millions of eggs, and they (or the subsequently hatched larvae) often become entrained in surface currents that disperse them—in some cases, all the way across ocean basins. Of course, most do not survive the voyage, but enough may do so to establish new populations, often in habitats very different from where they once lived, which given enough time may form new species. Sharks, on the other hand, have a few larger offspring that do not disperse like the bony fish. Most adult sharks are relatively small (less than 3.3 ft/1 m), so it would be risky to travel as far as bony fish eggs and larvae. Moreover, as many as 80 percent of shark species are benthic and have small home ranges as adults; they aren't dispersing either.

The diversity of bony fish species also ballooned due to the presence of an actual balloon, as bony fishes have an internal inflatable sac called a swim bladder, which they use to adjust their buoyancy. As a result, many bony fish species can hover almost wherever they choose in the water column. This adaptation allows them to occupy the abundant spaces in, say, coral reefs or mangrove systems. Sharks, which are heavier than water and must either rest on the seafloor or swim continuously, simply cannot exploit these habitats similarly. The swim bladder also facilitated the evolution of a small size in bony fishes which, again, gave them an advantage over sharks in certain environments.

An additional explanation for the high diversity of bony fishes is the complex bone structure of their heads. Whereas the skull, jaws, and adjacent structures of a shark's head involves only a handful of parts (a skull, upper and lower jaw, and a few supporting cartilages), the bony fish head is a complex jigsaw puzzle of dozens of bones. This provided natural selection with the raw material to produce a bewildering array of head shapes and mouth variations, and thus a diversity of feeding styles and opportunities to exploit new prey sources. Through this process new species developed.

Think of it this way: a kid can remain occupied for weeks building innumerable, highly creative structures with several dozen Lego bricks, but boredom would quickly ensue if the child had only a handful. Yes, the bony fish head is the anatomical equivalent of a set of Legos!

The functional significance of head shape is that no sharks protrude from burrows to gobble plankton like jawfishes, pick individual planktonic shrimp from the water column like the Mandarin Fish, shoot a stream of water at insects resting on terrestrial vegetation like an archerfish, eat detritus like the Striped Mullet, pick parasites like cleaner wrasses, or scrape algae from coral reefs like parrotfish. For that matter, no sharks are herbivores.

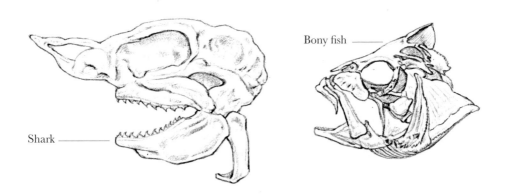

SKELETAL STRUCTURE
Simple (shark) vs complex (bony fish), skeletal differences that have allowed bony fish to evolve more varieties of head shapes and feeding styles than sharks.

Finally, since only four species of sharks are known to penetrate rivers and lakes, for reasons that we explain in Chapter 4 (see page 93), freshwater ecosystems represent a lost opportunity for further speciation. That audible sigh of relief you may be hearing comes from the Largemouth Bass, Bluegills, and carp from your nearby streams and lakes. Life is tough enough for them without worrying about sharks as well.

Do Kings Play Chess On Fine Glass Sets?

Or perhaps you prefer Drunken Kangaroos Punch Children On Family Game Shows. Still confused? Both of these silly sentences—and numerous others that often start with "Did King Philip"—are mnemonic sequences for remembering the hierarchy of Linnaean classification, based on the first letters of the Domain-Kingdom-Phylum-Class-Order-Family-Genus-Species system.

So, let's see where sharks are, taxonomically speaking, on the family tree we referred to earlier.

Domain Eukarya (Animals, plants, fungi, plus some single-celled organisms)

Kingdom Animalia (Organisms that do not photosynthesize or absorb their food)

Phylum Chordata (Animals with a skull and backbone, as well as around 3,000 kinds of oddball invertebrates like sea squirts that lack these)

At this point on the family tree, sharks share a branch with bony fish, other cartilaginous fishes, amphibians, reptiles, birds (or, more accurately, avian reptiles), and mammals (including you), as well as the jawless fishes and the extinct group of fishlike vertebrates known as placoderms that we mentioned in Chapter 2 (see page 42). To further distinguish sharks, we'll need to add subdivisions to the system we just introduced. Sorry, but we are unaware of mnemonic devices to recall these.

Subphylum Vertebrata (Organisms with a prominent skull, vertebral column, and brain)

Superclass Gnathostomata (All of the above minus the jawless fishes: the hagfish and lampreys)

The classification so far does not distinguish sharks from other cartilaginous fishes, tuna, alligators, frogs, seagulls, your pet cat and dog, or even humans. That will be next:

Class Chondrichthyes (The "cartilaginous fishes," including sharks, batoids, and chimaeras)

Subclass Elasmobranchii (The sharks and batoids)

If you recall, sharks and their cousins, the batoids (skates and rays) and chimaeras (ghost sharks), are all united as cartilaginous fishes in the class Chondrichthyes. All possess cartilage as their primary skeletal component, as well as the suite of other characteristics we introduced in Chapter 1 (see page 18). Within the Chondrichthyes are two subclasses: Holocephali (chimaeras, which mostly inhabit the deep sea) and Elasmobranchii, the most species-rich and diverse chondrichthyans. Elasmobranchs, as members of the subclass Elasmobranchii are commonly known, are further subdivided into three superorders:

Superorder Batoidea (skates and rays; sometimes called Batomorpha)

Superorder Squalomorphii (dogfish sharks)

Superorder Galeomorphii (galea sharks)

Batoidea includes more than 650 species of skates and rays. The Squalomorphii includes some 183 mostly cold-water sharks and the Galeomorphii a diverse array of 367 or more species of sharks. Since there are more similarities uniting the elasmobranchs than differences separating them, this book could easily have been titled: *Elasmobranchs: The Illustrated Biography.*

To visualize how different species are related to each other and how they evolved through time, taxonomists plot them all on phylogenetic trees. These diagrams depict the lines of evolutionary descent of different species and divide creatures into increasingly smaller groups based upon similarities in morphology or genetics. This means that when you look at a phylogenetic tree, species positioned on arms nearer to each other are more closely related and probably have more similarities.

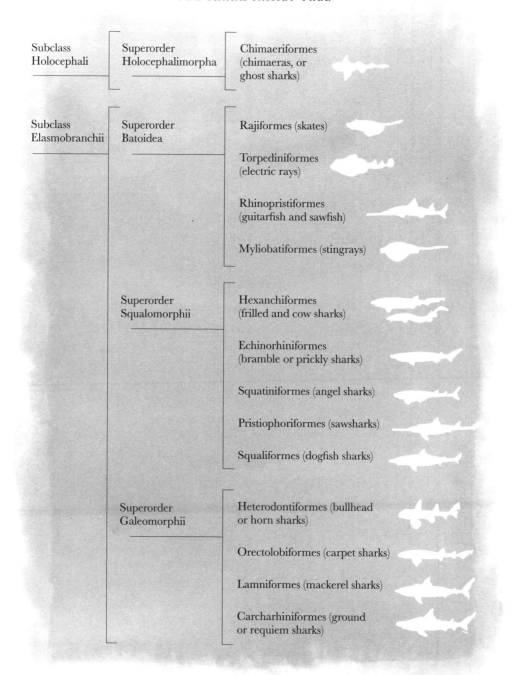

Subclass Holocephali	Superorder Holocephalimorpha	Chimaeriformes (chimaeras, or ghost sharks)
Subclass Elasmobranchii	Superorder Batoidea	Rajiformes (skates)
		Torpediniformes (electric rays)
		Rhinopristiformes (guitarfish and sawfish)
		Myliobatiformes (stingrays)
	Superorder Squalomorphii	Hexanchiformes (frilled and cow sharks)
		Echinorhiniformes (bramble or prickly sharks)
		Squatiniformes (angel sharks)
		Pristiophoriformes (sawsharks)
		Squaliformes (dogfish sharks)
	Superorder Galeomorphii	Heterodontiformes (bullhead or horn sharks)
		Orectolobiformes (carpet sharks)
		Lamniformes (mackerel sharks)
		Carcharhiniformes (ground or requiem sharks)

SHARKS AND THEIR CLOSEST RELATIVES

Sharks, skates, and rays are all united as elasmobranchs. Together with holocephalans
(ghost sharks), they constitute the 1,250 or so species in the class Chondrichthyes.

YOU TWO COULD BE SISTERS

The batoids and both shark superorders are sister groups—that is, they are each other's closest related group. As we learned in Chapter 2, the batoid lineage arose during the Late Triassic and Early Jurassic, between about 230 and 200 MYA, although some estimates go as far back as 266 MYA. Following the separation from the shark lineage, batoids evolved independently of sharks into the fascinating and diverse group that they are today. The superorder Batoidea comprises four orders: skates (order Rajiformes), electric rays (Torpediniformes), stingrays (Myliobatiformes), and guitarfishes and sawfishes (Rhinopristiformes).

The most easily discernible difference between sharks and batoids is their external form. The body plan of batoids consists of a flattened physique and the presence of "wings," which are actually extended pectoral fins fused along the entire length of the body. Exceptions to this generalization include an apparently dissident group of batoids, the guitarfishes and sawfishes, which "decided" to remain somewhat sharklike in their body forms as well as their ecological role. Not to worry, the surefire way to distinguish that Smalltooth Sawfish (*Pristis pectinata*) or Shark Ray (*Rhinus ancylostoma*) from the Sandbar Shark (*Carcharhinus plumbeus*) or Smallspotted Catshark (*Scyliorhinus canicula*) swimming in your favorite aquarium is the placement of their gill slits: in every single batoid the pairs of gill slits are ventral (on the underside), whereas they are lateral (on the sides) in all sharks.

We now arrive in the weeds, so to speak, distinguishing the two shark superorders from each other. We could devote the rest of the chapter, even the remainder of the book, to this task, but we won't. In the next few paragraphs, we'll limit ourselves to what we consider salient summaries of a sampling of the distinguishing features of the major groups of sharks, including our four focus species.

Squalomorph sharks

Sharks in the superorder Squalomorphii (from *squal*, meaning "dogfish") are a diverse group of about 183 species. These sharks are typically found in colder waters of mid- to high latitudes or in very deep water. To recognize each other in the near pitch-black, many can glow in the dark. Squalomorph sharks are characterized by the loss of an anal fin in all but one group (the frilled and cow sharks) and having only moderate jaw mobility. They are the most primitive of the two shark superorders, having arisen between 279 and 190 MYA, and are organized into five orders:

Hexanchiformes (frilled and cow sharks)
Echinorhiniformes (bramble or prickly sharks)
Pristiophoriformes (sawsharks)
Squaliformes (dogfish sharks)
Squatiniformes (angel sharks)

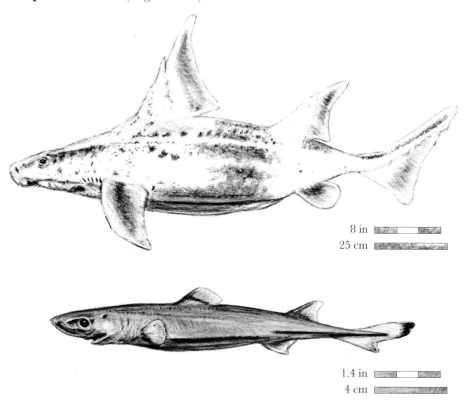

8 in
25 cm

1.4 in
4 cm

TWO SPECIES OF SQUALOMORPH SHARK
Two species contesting for the title of most beautiful shark, the Rough Shark
(top) and Dwarf Lanternshark (enlarged to showcase its magnificence).

Among the squalomorphs is the ancient lineage of Hexanchiformes, which all have more than five pairs of gill slits. This group includes the behemoth Bluntnose Sixgill Shark (*Hexanchus griseus*), which can reach 20 ft (6.1 m) in length, but there is nothing to fear, as they reside in the deep sea. Another abyssal critter is the slimy Gulper Shark (*Centrophorus granulosus*), which has big green eyes to see in the near pitch-darkness of the deep ocean. The Frilled Shark (*Chlamydoselachus anguineus*)—an eel-like shark with a reptilian-like head and pitchfork teeth—is also a part of this order.

The order Echinorhiniformes was differentiated relatively recently and contains only two species: the deep-sea Bramble Shark (*Echinorhinus brucus*), which has large, thornlike dermal denticles, and the Prickly Shark (*E. cookie*)—a sluggish, poorly known shark with large scales.

Unsurprisingly, the Spiny Dogfish (*Squalus acanthias*) belongs to the dogfish order (Squaliformes). This order is very diverse, ranging from the tiny Dwarf Lanternshark (*Etmopterus perryi*), 8 in (20 cm) in length, to the massive Greenland Shark (*Somniosus microcephalus*), at 23 ft (7 m), which can live for hundreds of years.

The Squalomorphs boast some of the weirdest and most wonderful of all the sharks. For example, the 23 species of angel sharks (order Squatiniformes) look more like batoids than true sharks and the remarkable sawsharks (order Pristiophoriformes) all sport a strange toothed rostrum that sticks out the front of their face.

Galeomorph sharks

In contrast to the squalomorphs, all the galeomorph sharks possess an anal fin as well as two well-developed dorsal fins that lack spines (except in the horn sharks). Evidence of galeomorph sharks in the fossil record dates to between 310 and 240 MYA, hence they are a more recent group than the squalomorphs. They mostly have highly protrusible jaws, an evolutionary advancement that enables a larger gape and a wider range of prey. While most species live in warmer climates, they are a diverse group, with about 367 species in as many as 24 families, separated into four orders:

Heterodontiformes (bullhead or horn sharks)
Orectolobiformes (carpet sharks)
Lamniformes (mackerel sharks)
Carcharhiniformes (ground or requiem sharks)

A major difference among the galeomorphs is the morphology of their snouts. One assemblage (around 50 species), including the horn sharks (family Heterodontidae), wobbegongs (family Orectolobidae), bamboo sharks (family Hemiscylliidae), nurse sharks (family Ginglymostomatidae), and the Whale Shark (*Rhincodon typus*), possess short snouts and no expanded rostral cartilage. The second group, which includes all of the requiem sharks (family Carcharhinidae) and the mackerel sharks, such as the White Shark, Basking Shark (*Cetorhinus maximus*), Sand Tiger (*Carcharias taurus*), Goblin Shark (*Mitsukurina owstoni*), the thresher sharks (family Alopiidae), and the mako sharks (*Isurus* species), all have elongated rostral cartilage and larger snouts. Additionally, only sharks in the orders Carcharhiniformes and Orectolobiformes have a nictitating membrane (or third eyelid). Also found in frogs, birds, and a few mammals, this clear, thin membrane slides across the eyeball to protect the eye.

Galeomorphs span the size spectrum, from small demersal (bottom-associated) sharks like the Smallspotted Catshark to the biggest fish in the ocean—the enormous Whale Shark. Despite reaching sizes up to 59 ft (18 m), Whale Sharks feed only on small plankton. Similarly, the second largest shark—the Basking Shark, which can reach 26 ft (7.9 m)—is also a filter-feeder.

Some galeomorphs, like Bull Sharks (*Carcharhinus leucas*) and Tiger Sharks, are voracious predators, while others, such as Nurse Sharks (*Ginglymostoma cirratum*), are more sedate and placid. These bottom-dwelling sharks use a suck-crush-spit-repeat feeding method on their hard-bodied prey (see page 92). The galeomorph group also includes the previously mentioned "walking shark"— the Halmahera Epaulette Shark (*Hemiscyllium halmahera*)—and Sand Tigers, who will feature prominently in Chapter 5.

Not to be outdone by the squalomorph gang, the galeomorphs also boast some truly spectacular oddballs, such as the Common Thresher Shark (*Alopias vulpinus*), which uses its long tail as a whip while hunting; the Goblin Shark, which can slingshot its entire jaw out of its mouth to increase its reach; and the flabby Megamouth Shark (*Megachasma pelagios*), a rarely encountered, deep-sea plankton-eater. Some of the strangest looking sharks in this group are the bull-head sharks (Heterodontiformes), such as the Port Jackson Shark (*Heterodontus portusjacksoni*), which are so named for their unmistakable, protruding brows. Believe it or not, these little sharks have been trained to recognize jazz music!

The galeomorphs also have some batoid-like sharks in their midst: the wob-begongs. While these sharks share the beautiful skin patterns of their carpet shark cousins (order Orectolobiformes), they look very different. With a flat-tened body and bizarre, undulating projections around their face that look like a beard, these sharks camouflage perfectly against rocks and corals, which allows them to ambush their prey.

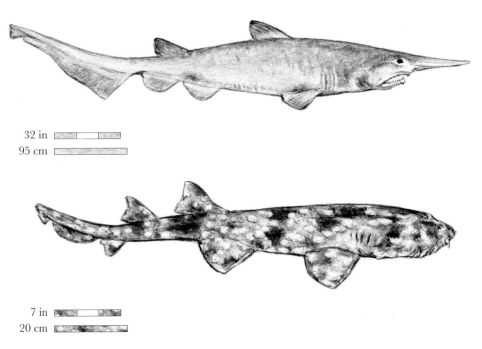

32 in
95 cm

7 in
20 cm

TWO SPECIES OF GALEOPMORPH SHARK
The Goblin Shark (top) and Ornate Wobbegong are more obscure
representatives of this group that includes the reef sharks.

70

The galeomorphs include by far the biggest order of all extant sharks: the Carcharhiniformes. This group is at least 270 species strong and includes some of the most iconic and recognizable species of sharks alive today: catsharks, hammerheads, and requiem sharks.

It's raining catsharks

The most species-rich group among the order Carcharhiniformes (and all shark families for that matter) is the catsharks—actually, three separate families: the catsharks (family Scyliorhinidae), finback catsharks (family Proscyllii-dae), and deepwater catsharks (family Pentanchidae). Of these three families, the most speciose group is the most recently created family of deepwater catsharks, which has as many as 114 species, at least for now. Our featured species, the Smallspotted Catshark, is just one of at least 170 different catshark species. Other representative species include the dazzling Chain Catshark (*Scyliorhinus retifer*), which is patterned with beautiful biofluorescent marbling across its skin, and the Pyjama Shark (*Poroderma africanum*), which was made famous after appearing as the villain in the fascinating documentary *My Octopus Teacher.*

As a group, catsharks are mostly small (less than around 3.3 ft/1 m) and found predominantly in cold and deep waters worldwide. Despite their diversity, you are unlikely to encounter many of these unless you are a deep-sea commercial fisher or scientist. These sharks lay eggs, have two small dorsal fins positioned far back on the body, and also have elongated, catlike eyes (hence the name).

Also within the order Carcharhiniformes is one of the most iconic groups of sharks on the planet, evoking reactions of *What? Otherworldly. Alien. Unreal. ...* the hammerheads. There are nine different species of hammerhead sharks. Apart from some obvious differences in size, each species can also be differentiated thanks to the varying size and shape of their iconic hammer-shaped cephalofoil. This appendage ranges from relatively small in the Bonnethead (*Sphyrna tiburo*) to the bizarrely wide head of the Winghead Shark (*Eusphyra blochii*)—an Indonesian species whose head is half as wide as its body length. In some species, the cephalofoils are smooth, while in others they are scalloped; some are straight, others rounded.

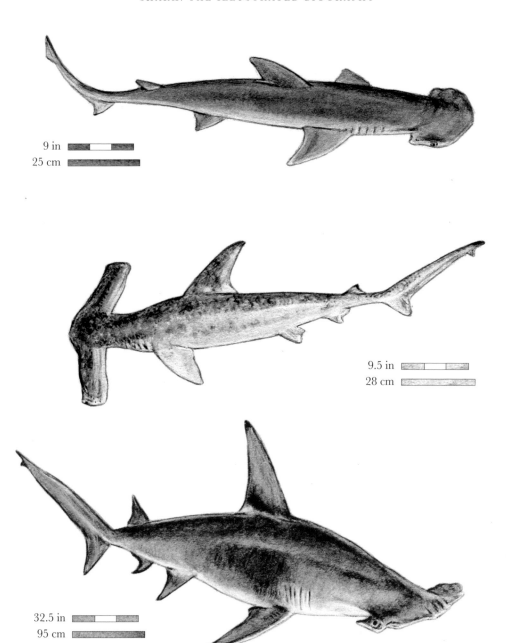

9 in
25 cm

9.5 in
28 cm

32.5 in
95 cm

THREE HAMMERHEAD SPECIES

Of the nine species, the Bonnethead (top) has the most petite cephalofoil and the Winghead
(middle) the longest, but the Great Hammerhead (bottom) reaches the biggest size.

Hammerheads are found in all tropical and warm temperate seas in both inshore and pelagic environments, with animals having been tracked from the surface to depths of nearly 3,280 ft (1,000 m). The smallest hammerhead species is the Bonnethead, which reaches only 4 ft (1.2 m). Where all hammerheads eat fish, invertebrates, batoids, and even other sharks, the Bonnethead may also eat and digest seagrass, making it one of only two species of sharks (the other being the Whale Shark) considered omnivores.

At the other end of the spectrum, the largest hammerhead is the aptly named Great Hammerhead (*Sphyrna mokarran*), which can reach a whopping 20 ft (6 m) in length. This elegant brute is easily distinguished from its hammerhead cousins by the absence of scalloping or a smooth curve on the leading edge of the head, and the presence of an enormous first dorsal fin, large pelvic fins, and a huge upper caudal fin. They are found throughout tropical and warm temperate seas in both inshore and pelagic environments, from the surface to 1,000 ft (300 m). As with other hammerheads, the cephalofoil plays a role in maneuverability and stability, but Great Hammerheads also use it to hunt; they have been witnessed using the hammer to pin rays to the seafloor before eating them.

Requiem for a dream

The stereotypical "sharkiest" of all the sharks can be found within the family Carcharhinidae—more commonly known as the requiem sharks. Numbering about 57 species, 34 of which are part of the largest shark genus, *Carcharhinus*, this group includes the Bull Shark, Lemon Shark (*Negaprion brevirostris*), reef sharks, and our Sandbar Shark, to name but a few.

Carcharhinids occupy all oceans, and many are potentially dangerous. Over 24 species are 5–16 ft (1.7–5 m) in length. Carcharhinids have a long, arched mouth with bladelike teeth, which are often broader in the upper jaw, a nictitating membrane, and no spiracles (with one exception). These are the dominant sharks in tropical and subtropical waters, and can be found in both inshore and pelagic waters from the surface to 2,625 ft (800 m). Four species are even able to tolerate fresh water. Bull Sharks, the best known of these, have remarkably been seen thousands of miles upstream and even left stranded in golf course lakes when flood waters recede!

A fascinating example from the carcharhinids is the Grey Reef Shark (*Carcharhinus amblyrhynchos*), which exhibits a fantastic suite of agonistic behaviors when it perceives something (or someone) encroaching on its patch. Arching the spine, dipping the pectoral fins, and swimming in an exaggerated, stylized manner, these sharks give ample warning to back off before they strike. While countershaded gray dominates in this family, there are some (very striking) exceptions. For example, the aptly named Lemon Shark has a subtle yellow tinge to its skin and there are no prizes for guessing the hue of the Blue Shark (*Prionace glauca*). One of the most abundant large (10.8 ft/3.3 m) sharks, found in the pelagic realm of the world's oceans, Blue Sharks are relatively slender and graceful, and have especially long pectoral fins.

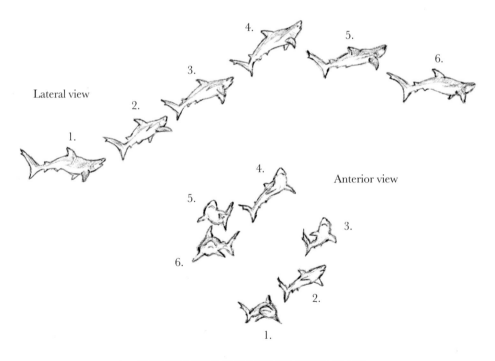

AGONISTIC DISPLAY OF THE GREY REEF SHARK
Grey Reef Sharks swim in exaggerated loops, with a hunched back and dipped pectoral fins, to warn off a threat.

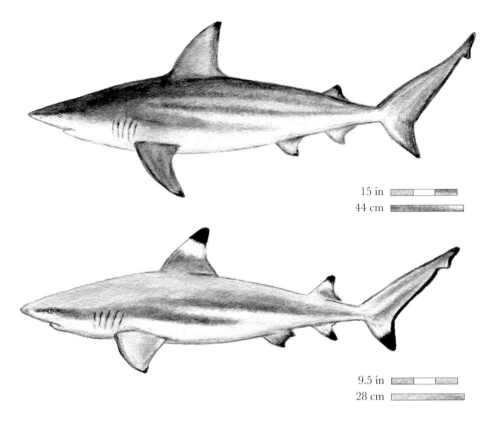

15 in
44 cm

9.5 in
28 cm

BLACKTIP AND BLACKTIP REEF SHARKS
*Blacktip Reef Sharks (bottom) can be identified by
their distinctive white stripe below the black tip on their
first dorsal fin. Blacktips (top) are noticeably larger.*

Colored fin tips are also a common feature in this group. For example, the Blacktip Shark (*Carcharhinus limbatus*) and Blacktip Reef Shark (*C. melanopterus*) both have (you guessed it) black fin tips. However, they are easily distinguished because the fins of the Blacktip Reefers look as if they have been dipped in India ink, and the species also has a distinctive white stripe under the black fin tip. Their lifestyles are also very different. Blacktip Reef Sharks live in groups within relatively small home ranges around coral reefs, whereas Blacktip Sharks are widely distributed both inshore and offshore and are frequently observed in dense, thousands-strong schools along their migratory pathways. Whitetip

Reef Sharks (*Triaenodon obesus*) and Oceanic Whitetip Sharks (*Carcharhinus longimanus*) have similar white markings. Whitetip Reef Sharks are small (5.2 ft/ 1.6 m), svelte sharks found around coral reefs, where Oceanic Whitetips are stocky, large (around 6.6 ft/ 2 m), pelagic sharks, found in temperate and tropical oceans. They are especially recognizable thanks to their uniquely paddle-shaped fins.

* * *

When many people hear the word "shark," the first thing that usually comes to mind is a large, streamlined, gray predator, but while this stereotype fits the carcharhinid sharks, it holds only for a very small subset of sharks as a whole. These archetypal sharks may populate the print, web, and TV media in endless repetition, and they are worthy of our fascination, but these beasts are atypical for two reasons. First, they are big, whereas the vast majority of sharks are actually very small (and pretty cute, if we may be so brazen!). Second, they live near to the surface or in shallow water, whereas most sharks actually live at dizzying depths—mostly below 660 ft (200 m). Throughout this chapter we have only very briefly scraped the surface of the diversity of sharks and, as technology develops and we are able to explore deeper than ever before, no doubt even more unknown, cryptic species will be discovered. So, while we join you in celebrating the White, Tiger, Bull, Blue, and Whale Shark, it is important, *critically important*, to tell the little guys' stories too. After all, everybody loves an underdog-fish.

$\overline{4}$

SHARKS IN
THEIR
HABITATS

TICKING ALL THE BOXES

A popular cable TV show follows prospective home buyers as a real estate agent presents several houses or condos for them to consider. Many of the clients are infuriatingly picky about what they want. Variously, they may desire an open floor plan or one more subdivided into walled rooms; a sprawling yard or a small courtyard; an expansive kitchen or a compact space; a cityscape or a more rural setting. What these potential buyers are doing, unbeknownst to them, is what ecologists would call habitat selection: determining some of the more significant environmental parameters of their living space.

How does a shark select its habitat? Are they as persnickety as these humans on the TV show, or does a shark simply remain in the general vicinity of where it was born? How far away does it roam? And if it roams, where does it go? What characteristics of its habitat are most important? In this chapter, we will explore the environmental features that determine where a shark lives and then examine these diverse habitats further.

Home is where the heart is

Surprisingly perhaps, sharks select their habitats in a similar way to humans. Substitute "open ocean" for "open floor plan" and "coral reef" for "subdivided into walled rooms," or "estuary" for "cityscape" and "deep sea" for "rural setting," and the parallels are vivid and unmistakable. Notwithstanding these similarities, there are differences in habitat selection between humans and other organisms like sharks. In essence, when selecting their habitat sharks reconnect with their evolutionary roots—glancing back in time and choosing (so to speak) the same or similar habitats as their most recent ancestors. Evolution is a powerful force, one that generally does not tinker with formulas that have been successful. So modern sharks, like the Sandbar Shark (*Carcharhinus plumbeus*) and Spiny Dogfish (*Squalus acanthias*), find comfort, as it were, living above the seafloor of continental shelves. Shortfin Makos (*Isurus oxyrinchus*) prefer residing in the upper layers of the

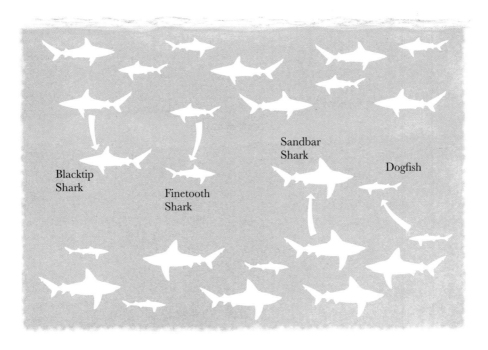

DIVIDING THE WATER COLUMN
*Sharks typically have specific habitat requirements, for example surface or
bottom, which may reduce competition.*

open ocean. These are the environments in which such species have lived for
generations: they are where their prey resides, where they've learned to evade
their predators, where they mate, have young, and so on.

It would be big news if an adult Shortfin Mako was observed swimming in
Boston Harbor, or a juvenile Sandbar Shark was seen from a submersible in the
deep sea. Similar events do happen, since given enough time even extremely
unlikely occurrences materialize and, when they do, they command media
space. For instance, in November 2022, a Blue Shark (*Prionace glauca*) was spotted in a narrow, marsh-lined creek in Cape Cod, Massachusetts. A woman
familiar with that salt marsh ecosystem initially assumed the creature was a
river otter, then identified the beast: "OK, well, that's not a river otter. It's got
fins." This led to a spate of reports featuring thrilling video of fins cutting
through the water.

Specific requirements

In general, where a shark lives is determined by its preferences and tolerances. First, there are abiotic (non-biological) environmental factors, including temperature, salinity, dissolved oxygen, pressure, current, light, and so on. In addition, there are also ecological considerations, like availability of food; competition for resources with other species (or even conspecifics; that is, members of their same species), or between different size classes (for example, juveniles and adults); and even risk of predation, since all but the largest sharks have their own predators.

These specifications are such that aquarists know they cannot simply throw different shark species into the large main tank but must provide the appropriate space, shelter, and conditions for each species. Many sharks do well in aquaria, such as Sandbar Sharks, Smallspotted Catsharks (*Scyliorhinus canicula*), Nurse Sharks (*Ginglymostoma cirratum*), and Sand Tigers (*Carcharias taurus*). Amazingly, Whale Sharks (*Rhincodon typus*), the largest living fish, have also been successfully displayed in aquaria. Some sharks, however, do poorly in captivity and habitat requirements in many cases provide an explanation.

In the 1980s, as a PhD student at Scripps Institution of Oceanography (SIO) in California, one of your authors worked with a team to help Sea World San Diego understand why White Sharks rarely survived in captivity for more than a few days. Fishing off the southern California coast, the aquarists captured a small (around 2.5 m/8.2 ft) juvenile White Shark (*Carcharodon carcharias*) and transported it to a small staging tank in a truck specially equipped to carry sharks, at which point it was observed and studied. Despite the precautions taken by the animal husbandry team, all of whom were knowledgeable and clearly had deep emotional connections to this White Shark, it died after 16 days in captivity. At the time, this was a record for the species. It was determined that the cause of death was inexorably tied to the biology of the species (it is a member of the same taxonomic family as the Shortfin Mako) and its habitat requirements. White Sharks are the elite athletes of the shark world and as such, even juveniles require very special handling and habitat requirements, especially space to roam orders of magnitude larger than the display tanks in marine aquaria. If we were to fill India's Narendra Modi Stadium—among the world's largest—with water instead of its usual 132,000 people, it would likely be too small for an adult White Shark.

This gorgeous shark specimen was also doomed by a combination of over-heating and a suite of stress-related responses that included a toxic buildup of chemical by-products of metabolism, water and salt imbalance, and other physiological dysfunctions, not to mention the alteration of the shark's behavior. In the White Shark's natural habitat, these harmful responses do not typically happen, but they occur during the incredibly stressful capture and transport process.

Paramount among the other habitat requirements for White Sharks is relatively cool water. While the body temperature of most sharks is the same as that of the water in which they reside, the White Shark, plus a handful of other sharks and bony fishes, is able to do something extremely rare in the marine environment (except for birds and mammals): trap some of the heat its body produces to elevate its own temperature above that of the surrounding water. We will discuss this remarkable adaptation further in Chapter 8 (see page 166). While this is beneficial for a White Shark in the wild, in the throes of capture stress, this heat accumulates and cannot be dissipated, causing what might be the shark equivalent of heatstroke, or a car overheating.

LOCATION, LOCATION, LOCATION

A mockumentary about sharks from the 1980s opens with: "Sharks are found in only two places—the Northern and Southern hemispheres." It is not hyperbole to assert that where there is enough water, there is shark habitat, although there are some qualifications. Sharks are found in marine environments from tropical coral reefs and mangrove systems to cold polar waters, from estuaries to the deep sea and open ocean. Perhaps surprisingly, a small number of species inhabit lakes and rivers. Big sharks, such as Lemon Sharks (*Negaprion brevirostris*) and Bull Sharks (*Carcharhinus leucas*), may occupy small creeks, while little sharks, like the lanternsharks (*Etmopterus* species), inhabit large spaces. Sharks like the Blue Shark, Oceanic Whitetip Shark (*Carcharhinus longimanus*), and Silky Shark (*Carcharhinus falciformis*) live in the open ocean's surface layer, whereas the Portuguese Dogfish (*Centroscymnus coelolepis*), which holds the depth record for sharks, has been caught at 12,000 ft (3,700 m).

The deep blue sea

We start with the ecosystem that, despite being the largest living space on the planet, is one you will never visit directly. And what a loss that is, for both you and the planet, since we often do not value what we cannot see. But if we could visit, we would find that the deep sea is endlessly fascinating, with a rogue's gallery of bizarre beasts (sharks included), the likes of which you might even call extraterrestrial.

The deep sea consists of those parts of the world's oceans deeper than 660 ft (200 m). Oceans cover over 70 percent of the Earth's surface, and most of this space, about 84 percent, is deeper than 6,500 ft (2,000 m), so we weren't exaggerating when we said this habitat is huge! The deep sea is shark central, with over half of all shark species residing there. In fact, the abundance and diversity of sharks peaks at depths of 1,300–2,600 ft (400–800 m). This degree of shark biodiversity is astonishing in light of the harshness of the deep-sea environment: it is a cold, dark, food desert of an environment. And consider the pressure—at 6,500 ft, the hydrostatic pressure is a bit like having the entire weight of a hippopotamus standing on your big toe. Ouch.

There is, however, one advantage to a deep-sea life: the suite of rather harsh environmental conditions is at least relatively constant. A little later we will see that most sharks of shallow coastal waters migrate in search of food, mates, and refuge. Since these aspects do not vary seasonally, deepwater sharks can breed year-round and there is no need to migrate. It's the ultimate staycation.

If a shark lives in the deep ocean, it still performs the same activities as a shark living in the shallows: locating prey, avoiding predators, and finding mates—in the dark, no less—while being mindful that conserving energy in such a food-poor environment is essential to survival. If one of your authors could use a metaphor that reveals his age, being a shark in the deep sea is like

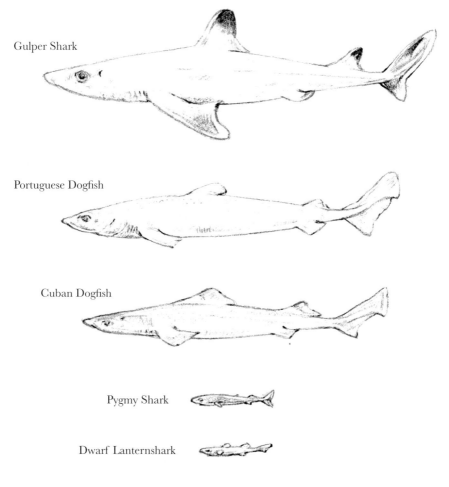

Gulper Shark

Portuguese Dogfish

Cuban Dogfish

Pygmy Shark

Dwarf Lanternshark

DEEP-SEA SHARKS
Five sharks that live deeper than 660 ft (200 m): Gulpers can grow to 5.7 ft (1.75 m), while the Dwarf Lanternshark stretches the tape measure to a mere 8.3 in (21 cm).

being Ginger Rogers—the less-heralded dance partner of the legendary hoofer Fred Astaire. She could, it is said, do the same amazing dance moves as Astaire, but she could do them backward and in high heels! Being a shark in the deep sea is, to say the least, challenging.

There is a depth limit for sharks. On average, the world's ocean is 13,100 ft (4,000 m) deep and plunges to a maximum of about 36,000 ft (11,000 m), yet sharks rarely occur deeper than 9,800 ft (3,000 m). If sharks can get to about 2 miles (3.2 km) deep, why not the full 6.8 miles (10.9 km)? Ecologically, there may not be enough food for sharks at the deeper depths. Physiologically, the high pressure may inhibit some of their core biochemical reactions, such as producing liver oil.

Deep-sea sharks differ in many ways from most of their shallow-water counterparts. Many deepwater shark species have very broad, sometimes global, distributions. This can be explained by the similarity of their environment over huge expanses—there is a general lack of barriers to their movement. That said, there is some diversity of habitats in the deep sea, including submarine canyons, seamounts, island slopes, and zones between the continental shelf and slope. These transitional zones and habitat edges promote biodiversity.

Inhabiting such a seemingly harsh environment requires adaptations and specializations that allow these sharks to survive and have offspring. The vast majority of deep-sea sharks have evolved to be relatively little, as it is much easier to sustain a small body in such a food-poor environment. For example, the smallest of all shark species, the Dwarf Lanternshark (*Etmopterus perryi*), would sit comfortably in the palm of your hand. Male and female Pygmy Sharks (*Euprotomicrus bispinatus*) reach only about 10 in (25 cm) and 8.7 in (22 cm), respectively, while Cookiecutter Sharks (*Isistius brasiliensis*), which were made famous in late 2023 by sinking an inflatable boat in the Coral Sea, are only 1.8 ft (55 cm) in length. We will talk in more detail about this fascinating species in Chapter 8.

At the other end of the size spectrum of deep-sea sharks is the largest shark you likely know very little about, if you've even heard of it. The Bluntnose Sixgill Shark (*Hexanchus griseus*) lives at depths of 1,000–3,300 ft (300–1,000 m) and can grow to over 15 ft (5 m), making it the ninth largest species of shark. As gigantism is such a rarity among deep-sea sharks, how then does the Bluntnose Sixgill defy the rules? First, it has large, green eyes, which are specialized for seeing in the blue-green light of its deep-sea environment. In addition, the Bluntnose Sixgill participates in a daily ritual known as a diel vertical migration. This is the largest animal migration on the planet, in which perhaps billions of tons of deep-sea organisms (small fishes, shrimp, squid, and other taxa) ascend from depths of around 3,300 ft (1,000 m) to around 660 ft (200 m) at sunset to feed on the plankton that have become concentrated there. This same group then descends to their deeper, food-poor waters at sunrise, completing the cycle. This migration may not match the drama of crocodiles ambushing wildebeest during their great migrations, but the interactions between predators and prey are equally fascinating and easily as ecologically vital. By participating in this daily habitat shift, the Bluntnose Sixgill avails itself of the caloric buffet available at shallower depths during the night.

Yet these adaptations are insufficient by themselves to account for the predatory prowess and size of this shark since large, green eyes and daily vertical migration are not unique among deepwater species. The Cookiecutter Shark, for example, also participates in this circadian yoyo swim. But here's where the Bluntnose Sixgill differs from most other deep-sea sharks: it forages on a mix of prey over a wide size range, such as crustaceans, cephalopods (squid and octopuses), bony fishes, other sharks, rays, and marine mammals, and it does so not only during its migration to shallow depths, but also on the seafloor, where it scavenges the carcasses of large animals like whales.

This shark species also has one additional trick up its sleeve. It has distinctive, wide, multicuspid teeth for sawing pieces of soft tissue out of carcasses. You might think such a big shark that can shear chunks of flesh from its prey would have jaws of steel. In fact, the opposite is true. Its jaws are poorly calcified and thus relatively weak and flexible. This means they bend across the body of large prey, so more of the sawlike teeth are in contact with the flesh at once.

Finally, why six gills slits when the typical shark configuration is only five? Just eight species of sharks, plus one batoid, have six paired gill slits, and two

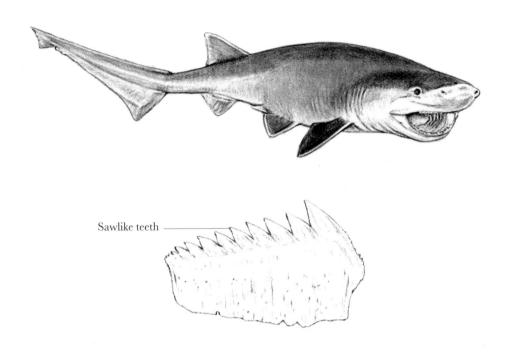

Sawlike teeth

GREEN-EYED BEAUTY OF THE DEEP
*The Bluntnose Sixgill, a massive (more than 15 ft/5 m), powerful
deep-sea shark, whose green eyes are specialized for the blue-green light
in its typical habitat. Note its distinctive teeth.*

sharks closely related to the Bluntnose Sixgill have seven. Since the core function of gills in sharks and batoids is for respiration, it might be reasonable to
hypothesize that the addition of one or two extra gill arches may reflect the
early evolution of these species in oxygen-deficient deep-sea waters.

Wide open spaces

Despite being continuous with the deep sea, open-ocean ecosystems—that part
of the ocean shallower than 660 ft (200 m)—are at the opposite end of the
shark-biodiversity spectrum. This area is called the photic zone, since it marks
the limit of sufficient light penetration for photosynthesis. Since there is no
barrier for movement between the open ocean and deep sea, many sharks trav-

el between the two to forage or rest, yet fewer than 25 shark species actually live in the open ocean. Whether due to TV documentaries or accounts from survivors of shipwrecks, you are probably familiar with several of these, particularly the Blue Shark, Oceanic Whitetip Shark, Common Thresher (*Alopias vulpinus*), and Shortfin Mako. Others, like the Longfin Mako (*Isurus paucus*), Bigeye Thresher (*Alopias superciliosus*), Silky Shark, and Porbeagle (*Lamna nasus*), might not be as well-known.

The relatively impoverished biodiversity of open-ocean sharks is readily explained. Firstly, there is a lack of abundant food, and less food translates into fewer kinds of large predatory sharks. Scarce food also means that oceanic sharks are generalists, so when they encounter a potential food item in their habitat, such as a sea turtle, bony fish, human, or even a suit of armor or a floating tire, one of their first, almost reflexive, reactions is to see if they can eat it. Secondly, since the open ocean is more or less featureless and monotonous—in other words, it has little habitat diversity—there are fewer ecological niches to fill and thus fewer kinds of sharks. What's more, the sharks of the open ocean have few barriers to movement and the space is expansive, so many open-ocean sharks—Blue Sharks and Oceanic Whitetips, for example—are far-ranging and exist as single populations. Since the creation of new species requires that populations of existing species somehow become separate and specialize, open-ocean sharks don't typically experience some of the common drivers of speciation.

Let's look more closely at two of the iconic open-ocean sharks, the Blue Shark and Oceanic Whitetip Shark. A strong argument could be made that these species are among the most elegant of sharks. With flanks of bright blue and a slender, 12.5-ft (3.8-m) torso, the Blue Shark is the runway model of oceanic sharks. If you see a shark on the open ocean, chances are it will be a Blue Shark, since it is the most abundant of the oceanic sharks and widely distributed. Equally graceful, albeit much stockier, is the Oceanic Whitetip, which reaches 11 ft (3.5 m). Both species, but especially the Oceanic Whitetip, share reputations as aggressive and potentially dangerous. To food generalists like these species, in an ecosystem where anything they encounter may be considered as prey, people represent nothing more or less than potential food that these sharks have been evolutionari-

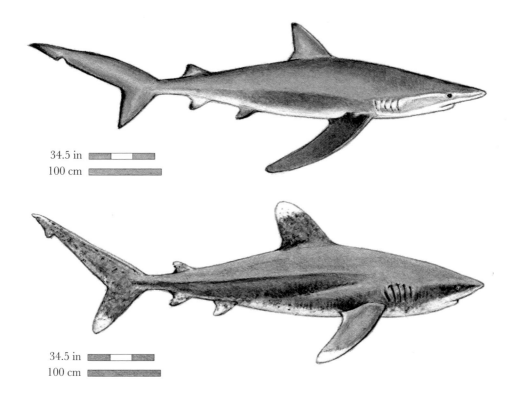

34.5 in
100 cm

34.5 in
100 cm

GRACEFUL OPEN OCEAN GLIDERS
*Two of the most common open ocean sharks, the Blue Shark (top)
and Oceanic Whitetip Shark. Both are food generalists, a necessity in
their food-poor habitat.*

ly programmed to investigate. Both species are thus naturally inquisitive and will
often circle and bump unfamiliar objects, including humans, apparently to identify both their palatability and any threat they might pose.

Coasting

We now move inland, to waters where you are much more likely to encounter
sharks, both because of their greater abundance and the frequency with which
we use these areas for recreation and fishing. In this section we will cover the
temperate and tropical shallows, areas of high shark biodiversity, as well as

places where the risk of extinction is heightened for more than three-quarters of the shark and ray species.

There are about 221,000 miles (356,000 km) of coastline in the world, and these ecosystems are home to about 40 percent of shark species, including all four of our featured sharks. Coastal habitats are an extremely diverse assemblage and vary widely in their environmental characteristics. Tropical ecosystems include coral reefs, mangrove systems, and seagrass beds, the last of which are also found in temperate shelf waters. Temperate coastlines also feature estuaries, kelp forests, and rocky, sandy, and muddy intertidal zones. A notable hotspot for temperate-water sharks is the coast of southern Africa. At any one time, between 15 and 22 species of sharks, including juvenile White Sharks, can be found in False Bay, a water body of 420 square miles (1,090 square kilometers) near Cape Town, South Africa.

The harshest coastal environment for sharks is, ironically, where most human-shark interactions (bites and attacks) occur—the intertidal zone, where the land is exposed and flooded daily, thanks to the tides. Beaches are high-energy, dynamic, and unstable environments with coarse sediments: a perfect storm of harsh factors that limits the abundance of both sharks and their prey. Schools of small fish, known as bait balls, will attract Blacktip Sharks (*Carcharhinus limbatus*), Sandbar Sharks, and others near to the beach, often creating exciting theatrical displays of huge splashes and charging shark bodies piercing the surface in pursuit of a meal.

Estuaries—where rivers meet the ocean—provide sharks with abundant food resources and protection from predation. This is especially important when these areas are used as nursery habitats, but we'll discuss this in more detail in Chapter 6 (see page 122). At the same time, the environmental conditions in estuaries can be stressful to sharks, as they consist of variably salty (brackish) water, where fresh water and saltwater meet and intermix, often on tidal cycles. Only about 50 shark species can withstand the (sometimes large) variations in temperature and salinity that characterize estuaries, and some of these species move away as their tolerance limits are reached. You are already familiar with many of the sharks found in estuaries, including Bull Sharks,

Sandbar Sharks, Lemon Sharks, Blacktip Sharks, Port Jackson Sharks (*Heterodontus portusjacksoni*), and Bonnetheads (*Sphyrna tiburo*). Others, like Finetooth Sharks (*Carcharhinus isodon*), Blacknose Sharks (*C. acronotus*), and Atlantic Sharpnose Sharks (*Rhizoprionodon terraenovae*), perhaps less so.

As an example of an estuarine shark, let's discuss a globally distributed species that is equally comfortable inshore and offshore in the upper 330 ft (100 m), on coral reefs and in mangrove swamps, and yes, in estuaries. We're talking about the Blacktip Shark, a bullet of a beast with a streamlined body, markings resembling a racing stripe, and, of course, its eponymous black-tipped fins (excepting the anal fin). Along the United States East Coast, the species is frequently observed in dense schools of thousands along its migratory route every spring and fall.

Blacktips, like all sharks, have an array of senses to locate their prey—primarily small fish and squid—and most of the time these senses work exquisitely, since they have evolved over a research-and-development period of about 450 million years. But their senses are occasionally fooled. In murky coastal waters, Blacktip Sharks sometimes apparently confuse human hands and feet with fish—and bite. Almost instantaneously, taste receptors in the mouth report the mistake to the brain, which replies to the jaws with a signal roughly translating to: "Not what I was expecting. Spit it out!" The evolutionary value of this reaction has little directly to do with humans, but is rather a generalized response to prevent swallowing prey that may be poisonous, of low nutritive value, too hard, or otherwise potentially harmful to the shark.

Coral corral

Coral reefs are special places, and they represent some of the most important real estate on the planet due to their extremely high productivity and structural complexity, which houses a wide variety of fish, invertebrate, and even reptile prey for sharks. Despite covering only about 0.2 percent of ocean floor or 96,000 square miles (250,000 square kilometers)—an area roughly the size of the state of Texas—reefs provide habitat for more than 130 species of sharks and batoids.

Some sharks occupy coral reefs or nearby waters more or less continuously, whereas others are transient visitors. Sharks on Indo-Pacific coral reefs include

Grey Reef Shark

Whitetip Reef Shark

Bull Shark

Blacktip Shark

Scalloped Hammerhead

Nurse Shark

CORAL REEF SHARKS
Coral reefs are critical habitats for many species of sharks.

the Grey Reef Shark (*Carcharhinus amblyrhynchos*), Whitetip Reef Shark (*Triaeno-don obesus*), Blacktip Reef Shark (*C. melanopterus*), which is not to be confused with the aforementioned Blacktip Shark, Tiger Shark (*Galeocerdo cuvier*), Silvertip Shark (*C. albimarginatus*), Galapagos Shark (*C. galapagensis*), Blacktip Shark, Scalloped Hammerhead (*Sphyrna lewini*), Great Hammerhead (*S. mokarran*), wobbegongs (family Orectolobidae), Tawny Nurse Shark (*Nebrius ferrugineus*), and Halmahera Epaulette Shark (*Hemiscyllium halmahera*). Sharks on Atlantic reefs include the Caribbean Reef Shark (*Carcharhinus perezi*), Tiger Shark, Bull Shark, Lemon Shark, Nurse Shark, and Great Hammerhead.

Here, we'll focus on two very different coral reef sharks: Tiger Sharks and Nurse Sharks. Both are widely recognized species and, depending on your perspective, also beloved, feared, or even reviled. Tigers have a blunted snout, teeth resembling a cockscomb, and eponymous markings, which are most vivid on juveniles. They are large sharks (16.5 ft/5 m) and found worldwide in tropical and temperate coastal waters. Nurse Sharks are chocolate brown, with two nearly equal-sized dorsal fins positioned far back on the body. Their mouth is near-terminal (as opposed to underslung), with long, protruding sensory barbels which are basically used to taste the water. There are four different species of nurse sharks in the family Ginglymostomatidae (one Atlantic, three Pacific), and they all are inshore, bottom-dwelling, largely nocturnal species, with a robust body that reaches about 10 ft (3 m).

Juvenile Tiger Sharks feed mostly on bony fishes. As adults, they also add sharks and rays, sea turtles, sea snakes, birds, mammals, and even invertebrates such as crustaceans and mollusks to their diet. Switching prey as they grow is common among sharks, especially for larger species, like White Sharks. You can identify dried Tiger Shark jaws both by their unique, can-opener-like teeth and their bent, deformed shape. You see, like the Bluntnose Sixgill Shark, surprisingly, Tiger Sharks have relatively weak jaws that are adapted for bending across the body of large prey (such as sea turtles and whale carcasses).

Nurse Sharks are durophagous—that is, they eat hard-bodied prey—and have strong jaw muscles. They employ a suck-crush-spit-repeat mode of feeding on bottom invertebrates (mostly mollusks and crustaceans). When a Nurse

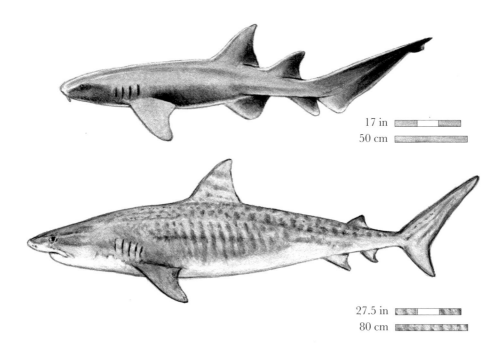

17 in
50 cm

27.5 in
80 cm

TWO CORAL REEF SHARKS
The Nurse Shark (top) is a mesopredator that dwells on the seafloor,
while the Tiger Shark is a true apex predator in the water column
of its coral reef habitat.

Shark encounters a Queen Conch, a large snail with a thick shell that deters any predators from the notion of crushing it, the shark positions its mouth over the snail's shell opening and applies suction strong enough to extract and inhale the soft animal from its protective shell—something you would be incapable of achieving by pulling with all your strength. It then proceeds to macerate the snail through a series of bite-and-spit maneuvers that result in sufficiently softened, bite-size morsels suitable for swallowing. If you've ever seen a Nurse Shark extracting a queen conch from its shell, you won't forget it.

Feeling fresh

Since liquid surface fresh water constitutes only 0.3 percent of the planet's water, it should come as no surprise that only about four shark species inhabit rivers and lakes. These include the Bull Shark (called the Zambesi Shark and Lake Nicaragua Shark in Africa and South America, respectively), plus about three species of river sharks found in the western Indo-Pacific, Papua New Guinea, and northern Australia: the Ganges Shark (*Glyphis gangeticus*), Northern River Shark (*G. garricki*), and Speartooth Shark (*G. glyphis*). About 50 species of batoids from three families are also capable of living in freshwater systems. Among these are the river rays (*Potamotrygon* species), a mostly inland, obligate freshwater assemblage in the Amazon basin, as well as the Largetooth Sawfish (*Pristis microdon*). A story published in 2023 recounts the saga of several Bull Sharks that had occupied a landlocked, largely freshwater golf course lake near Brisbane, Australia, from 1996 to 2013. Apparently, flooding of their nearby river residence allowed the juvenile Bull Sharks to penetrate the adjacent pond, where they became trapped once the floodwaters receded.

While numerous sharks inhabited freshwater habitats during the Golden Age of Sharks, since then they have evolved predominantly in marine ecosystems. The physiological and anatomical specializations required for sharks to inhabit

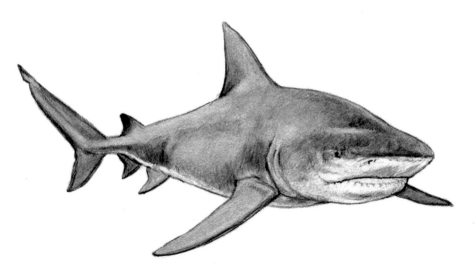

THE BULL SHARK
*A few species of sharks—most famously the Bull Shark—have a
wide salinity tolerance, so can survive in freshwater rivers and lakes,
as well as the ocean.*

freshwater ecosystems are so extreme that evolution did not favor them except in a few cases. Stop for a moment and consider what problems a shark or ray would face in fresh water. First, fresh waters have negligible salt content compared to the ocean, which presents great physiological challenges. As the homeostatic systems of sharks are exquisitely tuned to marine environments, in fresh water they would lose their own salts and take in water. Fresh water could also limit the ability of sharks to detect the electrical impulses emitted by their prey. In fresh water, sharks are even less buoyant than in salt water. To compensate, sharks would need to occupy a benthic niche or swim faster to keep from sinking, disadvantaging them energetically. Freshwater environments are often less thermally stable, so temperatures fluctuate much more, sperm are less viable in the water, which can have reproductive implications, and these areas may have only limited niche spaces for additional shark species—in other words, no vacancy.

Ice, ice, baby

Polar seas are broadly defined as those areas above latitudes of about 66°N or 66°S. These regions could be judged as the harshest environments sharks inhabit, based on the number of shark species found there. Only a handful of sharks are able to maintain physiological function at low temperatures and no shark species lives exclusively in polar waters. Sharks of the Arctic include the Greenland Shark (*Somniosus microcephalus*) and Pacific Sleeper Shark (*S. pacificus*), as well as the Porbeagle and Basking Shark (*Cetorhinus maximus*). The Southern Sleeper Shark (*Somniosus antarcticus*) and Porbeagle also occur in the polar waters of the Southern Hemisphere.

Winter sea-surface temperatures at these locations remain constantly cold—as low as the freezing point of seawater (28.5°F/−1.9°C) to about 35.5°F (2.0°C). These near-freezing water temperatures risk causing the blood of marine animals to freeze. Some polar bony fishes have evolved the protective measure of producing antifreeze proteins that lower the freezing point of their internal fluids, but similar adaptations have not been identified in sharks, except for a single report of antifreeze proteins in the skin of an unidentified shark purchased from a grocery store—hardly definitive verification. The freezing point of shark tissues

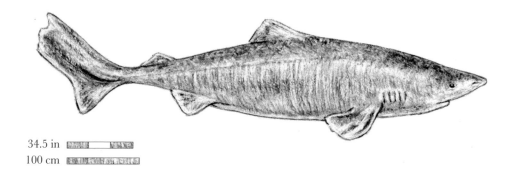

34.5 in

100 cm

THE GREENLAND SHARK
*The Greenland Shark, a denizen of cold, deep water in the North Atlantic
and Arctic, is a sluggish layabout of a shark that may live over 250 years!*

is naturally lower than that of bony fishes, since the "saltiness" (more specifically, the solute concentration) of the internal fluids of sharks is elevated over those of bony fishes, and the saltier the fluids, the lower the freezing point. This is why roads and driveways are salted in some areas amid icy conditions.

Whatever the underlying physiological mechanism that allows sharks to live in very cold waters, it is not surprising that most polar sharks are quite sluggish, since metabolic rate correlates with body temperature. The Greenland Shark, for instance, is the among the slowest of fishes, cruising about at just 0.76 mph (1.22 km/h). Compare this to the average 2–4 mph (3.2–6.4 km/h) walking speed of humans. Amazingly, though, this species is actually the largest high-latitude shark, growing up to 23 ft (7 m) or longer. Slow moving as it is, the Greenland Shark's diet consists of active prey like seals, whales, squid, and bony fish. Scientists have even found the remains of moose and polar bears in their digestive tracts. We must be sensible and presume that this was the results of scavenging dead bodies that fell into the water, but the imagination could certainly run wild picturing the low-octane chase scene of a lazy Greenland Shark versus a swimming polar bear, couldn't it?

5

SHARKS IN
THE WOMB

MINI ME

When the ovum, or egg, of a female shark unites with the male's sperm, a journey of unimaginable proportions begins. The voyage of development from zygote (fertilized egg) to adult shark is fraught with innumerable challenges for daily survival, with dangers lurking both inside and outside the womb. Yet it is also one of immense beauty and stunning diversity to anyone captivated by nature. Unlike bony fish, there is no larval stage and newborn sharks (called neonates) will emerge as near-perfect miniatures of the adults they will become if they survive. Sharks and their close relatives follow many of the same basic rules of development, although this does not equate to boring. On the contrary, evolution has prepared them for their journeys with an endlessly fascinating array of weird and wonderful adaptations.

To each their own

Even though the chondrichthyan fishes are the least biodiverse group of major extant vertebrate classes (exceeding only the jawless fishes in number of species), there is remarkable diversity in how they reproduce. Some species lay eggs; others give birth to live young. Some will nourish their offspring via a placenta, whereas others produce a kind of internal milk, with further variations among these variations. There is a remarkable range of shapes and sizes of egg cases, in how often females can breed, and even in sharks' reproductive anatomies.

If you compare the reproductive modes of sharks to those in other animals it becomes very apparent how astounding these are. For instance, the majority of bony fish species (around 97 percent) reproduce by spawning, with females laying minute eggs—as many as millions in the case of Cod or Bluefin Tuna—and males subsequently fertilizing these eggs in the water. The developing embryos within these eggs are nourished exclusively by resources stored in the yolk and they then hatch out into the world as larvae, typically very small larvae. Eventually they morph into their adult shape—if, that is, they survive all the threats being so tiny exposes them to. This strategy is conserved quite

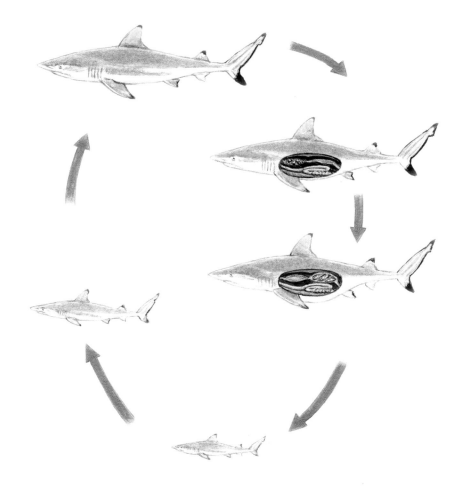

LIFE CYCLE OF THE BLACKTIP SHARK
*Clockwise from upper left: Adult female Blacktip Shark; female showing
ovary containing mature ova; pregnant female with three embryos in
crypts; young-of-year, juvenile. Drawn approximately to scale.*

broadly throughout the bony fish world. Consider too the very different strategy of the mammals, which all use internal fertilization and give birth to a few
live, but typically larger, young, each with a higher chance of survival.

Pretty much the only reproductive trait conserved across all sharks is sexual
reproduction (just to be truthful, there are a few rare exceptions, but we will defer

discussion on these for now), in which a male and female combine their genetic material. This mixing of heritages means that all offspring are unique individuals and not genetic duplicates of their parents, endowing some with traits that enhance their survival and others that make them somewhat less fit. Genetic variation is the raw material of natural selection and accounts for the diversity we see in sharks and their relatives, and in all sexually-reproducing organisms.

Sexual reproduction in sharks involves internal fertilization, in which the male and female copulate and the gametes (sperm and ova, respectively) are conjoined inside the mother. Compared to spawning, fertilizing the eggs internally lowers the offspring's risk of mortality as they are housed and at least partially develop safely inside the female. Since sharks are predators, typically in the top levels of the food web, the extra measure of protection inside a creature that is more of a threat to other organisms than vice versa should not be discounted. Sharks have the luxury, so to speak, of heavily investing resources in a small number of thoroughly nourished and well-developed offspring, which are more likely to make it to adulthood. The vanishingly small Cod or Bluefin Tuna hatchling is an easy menu item compared to, say, a Sand Tiger (*Carcharias taurus*) newborn. Which hatchling would you bet on to survive? Of course, as we will see later, the advantage of having fewer, larger offspring becomes a major liability when populations are depleted by overfishing and other human impacts, and it imposes real stresses on the female, who must carry her young as they develop.

Boy meets girl

The act of copulation in sharks is similar to that seen in humans, at least superficially, and this leads to a common question: How do you tell the difference between male and female sharks? If you recall, male sharks have two (yes, two) very distinctive external sexual organs on their underside. These claspers are quite conspicuous, especially as the male grows to maturity and they become increasingly large and stiffened. While they basically have the same function as the human penis, a shark's claspers are not an independent appendage but are rather modified, grooved, tubular parts of the paired pelvic fins.

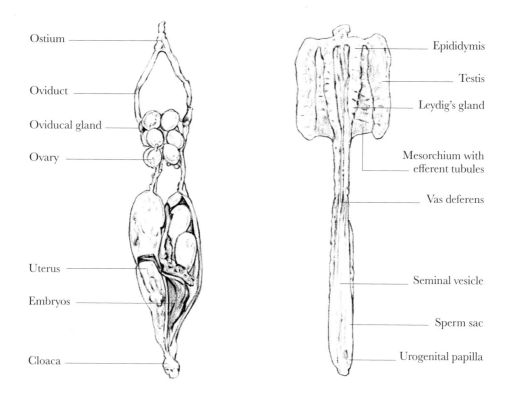

Ostium

Oviduct

Oviducal gland

Ovary

Uterus

Embryos

Cloaca

Epididymis

Testis

Leydig's gland

Mesorchium with
efferent tubules

Vas deferens

Seminal vesicle

Sperm sac

Urogenital papilla

SHARK REPRODUCTIVE ORGANS
Anatomy of the reproductive organs of female (left)
and male (right) sharks.

Although both are sperm-transfer organs, claspers and the human penis also differ in how they function. Firstly, claspers do not become engorged with blood and, secondly, sharks do not ejaculate. These factors are both due to the presence in males of an onboard, short-term seawater storage organ called a siphon sac. This sac, which can be as long as a third of the shark's body, sits just under the skin of its belly. Its function is to supply one of the claspers with pressurized water. Once copulation begins, a single clasper is rotated between 90 and 180 degrees and inserted into the female's cloaca—the common opening into which the genital, digestive, and urinary tracts empty. Next, two separate events must occur. First, sperm must make their way from storage in the sperm sac through a pore (called the urogenital papilla) into the now rotated clasper.

Then, often with impressive and enthusiastic thrusting, the male shark will bend his body, compressing the saltwater-filled pouch and causing it to empty its contents into the clasper through a different pore, the apopyle. This seawater-semen concoction will then be propelled into the female via an external groove in the clasper. The shape and size of claspers vary enormously between species, and the tip of the clasper can actually splay out and project some truly bizarre structures. These protrusions serve to anchor the clasper in place during vigorous copulation. They can be so effective that sometimes the female must shake vigorously to dislodge the male after sex!

Differences between shark and human reproduction also extend to how the sperm are produced and stored. While sharks may make their sperm in a pair of testes as humans do, these organs are inside the body and closer to the shark's head than to his claspers. After being produced, the semen is propelled along some internal ducting, where conditions allow the sperm to be nourished to maturity, and moves into a sac called the seminal vesicle before being transferred to the final storage destination, the sperm sac. This latter structure is also known as the alkaline gland, since it adds a basic (an alkaline or high pH) fluid. Since the sperm are deposited in a region that might expose them to the female's acidic urine, this alkaline bath may help to protect them.

The female form

The internal anatomy of female sharks also has some amazing and startling differences compared to the female human body. Most notably, female sharks have two reproductive tracts, including two ovaries (where eggs are made) and two uteri (where embryos develop). In many of the carcharhiniform sharks (the requiem, or ground sharks), the female's reproductive tracts are both functional at the same time, so eggs can be fertilized and develop in each uterus simultaneously after one mating. In other species, only one of the two ovaries will be functional at once, but will supply viable eggs to both of the uteri. Yet other species have only a single functional ovary, as the other has atrophied and become useless. In a mature female shark, the ovaries are large and unmistakable; the bright yellow, globular follicles containing the budding eggs occupy a huge proportion of the anterior body cavity. The size and functionality of the uteri, however, can vary between species, depending on how their young develop (more on this later).

Some female sharks also have specialized tubules within their reproductive tract where they can save viable sperm: a veritable sperm bank. This means the female can retain sperm in her oviducal gland for an extended duration after mating, releasing the sperm to fertilize her eggs only when she is ready. Lemon Sharks (*Negaprion brevirostris*) were the first species of shark for which sperm storage was documented, but we now know it is quite common in sharks. Chain Catsharks (*Scylliorhinus retifer*) kept in captivity have been known to become pregnant from sperm stored seven years after they were last housed with a male. Female Gummy Sharks (*Mustelus antarcticus*) will even mate with males before these females are sexually mature—that is to say, before their bodies have even developed enough to have offspring. The female will retain the sperm until she is sufficiently sexually developed to become pregnant. Quite the party trick.

LOVE BITES

Since sharks live in environments that challenge our ability to study them, we lack knowledge of many aspects of their biology. One facet which remains ever mysterious is how they mate. It is rare to witness shark courtship, and there are only a handful of species of sharks in which scientists have actually observed mating. For the vast majority, we know nothing about where and when they breed, and while we suspect sharks might be able to smell viable mates, we know very little about how they find or choose their partners. We do know that sharks do not mate for life like penguins or wolves, for example, and that the males do not offer nuptial gifts in the form of a nest or food. Instead, sharks select a different partner each time they are fertile. If the female is receptive, they mate, and then the two part, presumably never to see each other again— the shark equivalent of a one-night stand.

Where scientists have been lucky enough to witness that special moment when sharks are mating, they have learned that shark sex is brief—usually between 15 and 20 minutes—and it is rough. Very rough! In many species of sharks the male will swim alongside or behind the female before commencing copulation. He can approach from either side, using his left clasper to mate if on her left or his right when on her right. As mating begins, the male shark will often bite on the female's pectoral fin to keep her in position while he inserts his clasper. This seems to be a shark's international flirting signal, as it is a behavior we see across many different types of distantly related sharks, from Bamboo Sharks (*Hemiscyllium freycineti*) to Blacktip Reef Sharks (*Carcharhinus melanopterus*). Yet these bites can be significant, leading to scars on the female's flanks. Female sharks have evolved thicker skin on their sides, fins, and around their gills to protect them from these "love bites."

In fact, mating can be so dangerous to females that many species of sharks do not live in mixed sex groups (a practice known as sexual segregation). Instead, the females will actively avoid the males to eliminate, or at least limit, sexual harassment. For example, female Smallspotted Catsharks (*Scyliorhinus canicula*) all refuge together in closely bonded social groups within labyrinthine caves, while the males live in deeper offshore waters. This drive to evade the males is so strong that female sharks will even live in areas outside their optimal thermal tolerance—waters that are too hot or too cold—potentially putting themselves under physiological stress, just to avoid mating.

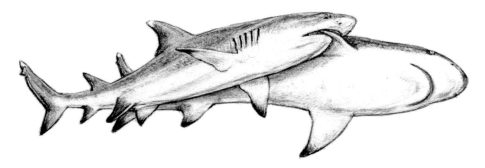

TOUGH LOVE
In many species of sharks, including these Blacktip Reef Sharks, the male bites down on the female's pectoral fin during mating.

Kama sharkra

Scalloped Hammerheads (*Sphyrna lewini*) display an especially unique and remarkable mating behavior—an extended copulatory freefall. After the biting, thrusting, and thrashing part of proceedings is completed, Scalloped Hammerheads will suddenly stop swimming as they are mating. Entwined together, they will freefall from high in the water column all the way down to the ocean floor, as far as 130 ft (40 m), or even deeper in the open ocean. After landing, the sharks will remain motionless on the sand for a while before separating. The male will then follow the female around for a few minutes before each goes their separate ways. Scientists wonder if this behavior allows the male to ensure no other male gets to his female, increasing his likelihood of reproductive success, but we do not know for sure.

Catsharks (family Scyliorhinidae) are quite the contortionists during mating. The male will bend his body all the way back on itself, curling himself around the female and almost completely enveloping her. He's so twisted up, he assumes the shape of a donut as they mate. Other species of sharks have sex in groups (known as group copulation). For instance, in Whitetip Reef Sharks (*Triaenodon obesus*) and Nurse Sharks (*Ginglymostoma cirratum*), multiple males will all pursue one female in a frenzied ball of bodies and fins. It might sound quite violent, but in fact, in these situations it seems the girls have all the power. Rather than being molested by multiple males, the female shark will choose a specific male out of the group, positioning her body in such a way that she can reject unfavorable males and acquiesce only to the mate of her choice within the fray. The males are forced to compete for her affections and only the best will be successful—maybe the biggest, the strongest swimmer, or the most agile. Who knows what is alluring to a female Whitetip Reef Shark? This female mate choice allows the mother to ensure her young are produced from only the most exceptional genes, from the finest specimen she can find, thus ensuring they will become strong, sexy, and successful themselves.

On the other hand, females of some shark species will choose to mate with multiple different males. In fact, some females can even have one litter of pups with numerous fathers, so some of the offspring in the litter are half-siblings. This is known as polyandry, or multiple paternity. This happens when a fertile female mates with several partners over a short time and her eggs are fertilized by different males. Multiple paternity is found throughout the animal kingdom—in mice, birds, and turtles, to name but a few—and it is widespread across many different groups of sharks, including Lemon Sharks, Bull Sharks (*Carcharhinus leucas*), some of the hammerheads (family Sphyrnidae), reef sharks (*Carcharhinus* species), and smoothhounds (*Mustelus* species). While relatively rare in Spiny Dogfish (*Squalus acanthias*), where genetic studies have taught us that only about 17 percent of egg clutches come from multiple fathers, polyandry can be very common in some species—as high as 100 percent of litters in Shortfin Mako (*Isurus oxyrinchus*), for instance.

Like a virgin

At the absolute opposite end of the spectrum from polyandry is an incredible phenomenon which has been documented only in a very small number of shark species: virgin births. Known as parthenogenesis, this reproductive marvel involves a female shark producing a litter of offspring without ever mating with a male. Surprisingly, parthenogenesis is not uncommon in other animals, like reptiles. The most famous example in sharks is that of a captive Zebra Shark (*Stegostoma tigrinum*), which gave birth to litters of fertile pups despite never having been kept in a tank with a male. It has also been documented to happen very rarely in Bonnethead Sharks (*Sphyrna tiburo*), Blacktip Sharks (*Carcharhinus limbatus*), Whitespotted Bamboo Sharks (*Chiloscyllium plagiosum*), and Whitetip Reef Sharks.

It is possible for some female sharks to breed without paternal DNA because of the presence of a polar cell, which basically assumes the function of a sperm. This polar cell, with half the genetic complement of a fertilized egg, is normally inactive, arising as a part of the egg production process. However, in parthenogenesis, the polar cell fuses with the mother's egg to create a viable cell; that is, with its complete complement of genetic material, half of which would otherwise have been provided by the male.

It's thought that some sharks can switch to this method of asexual reproduction when there is no male around to breed with—when they are housed in an all-female tank, or when their populations in the wild are so fragmented that they rarely run into a viable male, for instance. The benefit is that it allows the female to continue to breed, passing on her genes and bolstering the population size. The downside to parthenogenesis is that it produces only female pups with half the amount of genetic variation that would be present if they were produced through sexual reproduction. This can lead to serious genetic bottlenecks in subsequent generations, and the population can over time become inbred and particularly susceptible to environmental disruptions or diseases.

DON'T PUT ALL YOUR EGGS IN ONE BASKET

Sharks reproduce in a myriad of ways, a phenomenon that has allowed them to evolve to fill many different evolutionary niches, and to breed successfully in various situations. Rather than separate categories, there is more of a continuum of different ways that sharks reproduce. In order to simplify everything, we can categorize shark reproduction in two broad groups: those that lay eggs (oviparous sharks) and those that give birth to live young (viviparous sharks).

Egg laying

Let's start with the egg layers. Sharks that lay eggs are said to be oviparous, which literally translates from the Latin for "to bring forth an egg." Oviparity involves a female shark laying fertile eggs after she has mated, just as in birds and reptiles. After the eggs have been deposited in a safe spot on the ocean floor (not adrift in the water column), neither parent protects nor sustains the eggs. Instead, the growing embryos utilize the fatty, proteinaceous yolk deposited by the mother until they have developed sufficiently to be able to hatch. Since the only nourishment the growing embryo can access is in the egg, as the shark matures prior to hatching it loses weight, since metabolism is not 100 percent efficient and, yes, the embryo poops and pees.

Around 40 percent of all species of sharks are oviparous, with egg laying occurring across many different taxonomic shark groups. Most oviparous sharks are smaller species that live on or near the ocean floor, like the catsharks and their batoid cousins, the skates. Egg laying even pops up randomly among other reproductive strategies within groups of closely related species. All members of the order of bullhead sharks (Heterodontiformes) are oviparous, but egg laying can also be found in some species of carpet sharks (order Orectolobiformes) and ground sharks (order Carcharhiniformes), which are separated by millions of years of evolution.

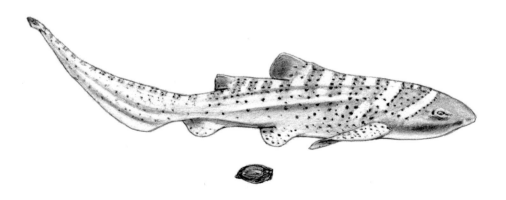

A ZEBRA SHARK AND HER EGG
Zebra Sharks' egg cases are the largest of any oviparous shark species, reaching about 6 percent of their body length.

In general, oviparous sharks can produce eggs year-round, but there are often seasonal breeding periods lasting several months where we see the majority of females laying eggs. Zebra Sharks have the largest egg case of all sharks. Despite reaching a maximum size of only about 8 ft (2.5 m) from nose to tail, Zebra Sharks lay eggs which are around 7 in (17 cm) long, which is impressive.

The award for the most visually striking shark eggs probably goes to the Horn Shark (*Heterodontus francisci*) and its bullhead shark relatives. These darkly colored, almost black, eggs are shaped like fattened corkscrews. The threading allows the eggs to be effectively screwed into crevices in rocks, where they can be safe from harm. A female Horn Shark will even carry the eggs in her mouth until she finds the right spot to stash her unborn young.

The Smallspotted Catshark is a shallow-water egg layer. Females deposit a pair of eggs at a time, one from each oviduct, reaching a total of around 60 (range 40–240) per year. The tough egg cases are about 1.6–2.4 in (4–6 cm) long, with curly, wiry tendrils, which become wrapped around algae and other structures on the substrate. This holds the egg case in place for the eight to nine months or so it takes the pup to develop and hatch.

After shark eggs have been laid and while they continue to develop outside the mother, they are especially vulnerable. Some oviparous shark embryos have evolved a particularly nifty strategy to survive—they can sense predators before they have even hatched. Brown-banded Bamboo Sharks (*Chiloscyllium punctatum*) develop for around five months after being laid before hatching. During the later stages, slits forms in the egg case to allow water to circulate, providing oxygen and eliminating wastes. This is a dangerous time because the scent of the pups can waft into the surrounding water. But natural selection comes to the rescue. The electrosensory system of the tiny sharks detects the minute electrical charges generated by a potential predator's muscle contractions as it swims close by. Whereas the unborn sharks normally wriggle around quite joyfully inside their egg cases, which prevents their surroundings dangerously stagnating, when they detect a prospective predator, their movements cease—they don't even pump water over their gills, so are effectively holding their breath—until the threat has passed.

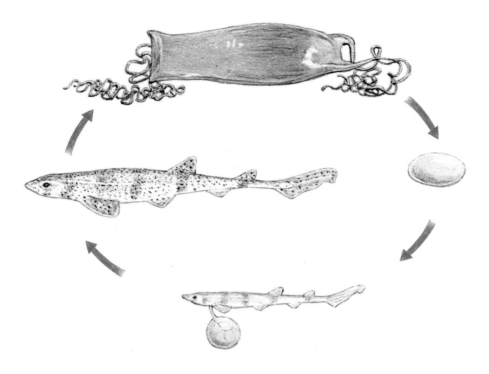

DEVELOPMENT OF A SMALLSPOTTED CATSHARK EMBRYO
*Clockwise from the top: the hardened egg case; the fertilized embryo inside
the case; the proteinaceous, fatty yolk attached to the embryo; the neonate.
When yolk reserves are exhausted after 6–9 months, the shark hatches
from its egg case and must begin foraging for food.*

Live birth

Numerous sharks give birth to live young, a mode known as viviparity. In these
species, the fertilized eggs do not form a hardened shell after mating and,
instead of being laid, they develop inside the mother with only a thin membrane around them, which is lost before they are born alive.

Like oviparous sharks, all viviparous species have a yolk for sustenance during
the early stages of pregnancy, but there is then a wide spectrum of methods for
how the mother subsequently feeds her young as they develop. For some species,
after they have broken out of their egg cases, the embryos continue to use yolk
throughout the rest of their development while free in the uterus. This is known

as lecithotrophic viviparity and is found in about a quarter of all living sharks, including both species of frilled sharks (*Chlamydoselachus* species). Female frilled sharks will be gravid for an estimated three and a half years before giving birth to their young, quite possibly the longest gestation among animals.

In other species, the pups will switch to a different form of nourishment after the yolk has been absorbed and, once again, there are a myriad of different strategies between species. These can be broadly grouped into species that feed their embryos via a placenta (aptly named placental viviparity) and those that do not (this is known as yolk-sac viviparity but also aplacental viviparity or ovoviviparity).

Around 40 percent of sharks, which includes around 160 species of the superorder Squalomorphii (the dogfish sharks), utilize yolk-sac viviparity, and this strategy is also common in rays. Like other reproductive methods, yolk-sac viviparity can be found across many different groups of sharks, from the angel sharks (order Squatiniformes) to the Tiger Shark (*Galeocerdo cuvier*), and it also pops up randomly within families. The Spiny Dogfish is yolk-sac viviparous. Instead of a tough shell, a thin membranous envelope is deposited around groups of fertilized eggs, one to four per envelope, forming a structure known as a candle. These candles move into the two uteri and serve to maintain the intrauterine conditions required for the embryos to develop. After breaking out of this membrane about halfway through development, the embryos are then free in the uterus and nourished by what remains of their external yolk sac. At 22–24 months, the gestation period of the Spiny Dogfish is among the longest known. Young, typically numbering six to seven, but reaching as many as 15, are born headfirst at 8–13 in (20–33 cm), with the spines encased in a sheath, which protects the mother from injury.

Similarly, all ten species of sawsharks (order Pristiophoriformes) are yolk-sac viviparous. You might wonder how on earth the female can carry pups armed with a tooth-covered rostrum (or saw, hence the name) projecting from the front of their face. During pregnancy a sawshark pup's rostral teeth will be folded back against the side of the saw, so they do not harm the mother. They will then pop out into their proper position after birth.

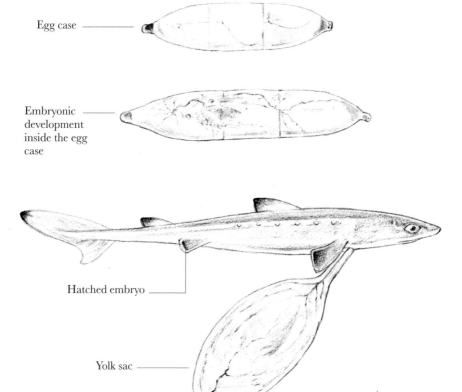

Egg case

Embryonic
development
inside the egg
case

Hatched embryo

Yolk sac

REPRODUCTION IN THE SPINY DOGFISH
*Spiny Dogfish pups first develop within an egg case, and then hatch
out—with their yolk sac still attached—to continue their development
inside the mother before birth.*

Yolk-sac viviparity is so common because it balances some of the pros and
cons associated with the more extreme reproductive methods on the spectrum;
the cost to the mother is relatively low compared to placental viviparity, but the
pups are protected inside the female for a longer period compared to oviparity.
Yet the costs to the female can vary depending on the type of nutrition she
subsequently provides for her developing pups.

Tiger Sharks utilize a strategy of intrauterine nourishment not found in any
other species of shark. After the yolk has been depleted, the embryos switch to

feeding on a nutritious fluid in the womb, a process known as embryotrophy. A female Tiger Shark can produce a litter ranging from 45 to 60 pups from one pregnancy and each pup will see as much as a 1,000 percent weight gain throughout the pregnancy, one of the highest increases of body mass for a developing embryo found in any species of shark.

The gestation period for the White Shark (*Carcharodon carcharias*) is about 11–12 months. Females are thought to give birth every two to three years and the off-spring are born when they are 3.3–5 ft (1–1.5 m). The number of pups is not known with certainty, but the record is 14 in a single litter. You would be forgiven for thinking that the predatory process of the White Shark exists only after birth, but you'd be wrong! The White Shark and its four mackerel shark cousins—the Shortfin Mako, Longfin Mako (*Isurus paucus*), Salmon Shark (*Lamna ditropis*), and Porbeagle (*Lamna nasus*)—and the false catsharks (family Pseudotriakidae) prac-tice a form of intrauterine predation called oophagy (meaning "egg nourish-ment"). When most of the yolk has largely been absorbed by the embryos, the mother begins ovulating unfertilized ova. The embryos swim within the uteri and

WHITE SHARK EMBRYO
Before birth, White Shark embryos gorge themselves on
unfertilized eggs produced by the mother, to such an extreme
that they develop pouchlike stomachs.

113

become predators-in-training, using their precocious, sharp teeth to consume the new ova at great speed. And if you also thought that White Sharks could not be cute, then wrong again. Like chipmunks overstuffing their cheeks with acorns, the White Shark embryos pack their tummies with so many nutritious unfertilized ova that they develop egg stomachs. Recent evidence suggests that the uteri also secrete a nutritive fluid early in the development of White Sharks and their relatives in the order Lamniformes. This results in a greater than tenfold increase in weight gain before birth.

Even more bizarre and arguably the most morbidly fascinating example of yolk-sac viviparity is that of the Sand Tiger, the embryos of which prey on their brothers and sisters in the womb. This is known as intrauterine cannibalism, or adelphophagy. While the female may harbor as many as 50 fertilized eggs after mating with multiple males, by the time she gives birth, she will produce only one pup per uterus—the strongest two. It might sound barbaric and cruel, but this strategy has developed because it benefits the mother to give birth only to the strongest young. It's survival of the fittest in a dog-eat-dog (or shark-eat-shark) world.

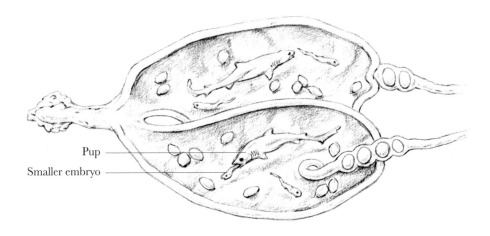

Pup
Smaller embryo

INTRAUTERINE CANNIBALISM
Sand Tigers have a unique reproductive method, whereby the two most advanced unborn pups will kill and eat all their siblings in their respective uteri before they are born.

Placental structure

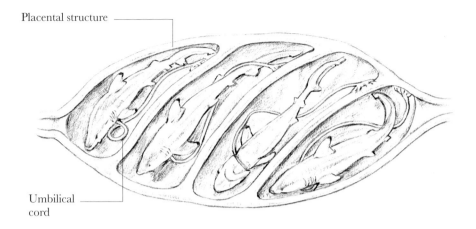

Umbilical
cord

PLACENTAL NUTRITION
*In placental viviparous species of sharks, each unborn pup is nourished
via a placental attachment that connects to their mother.*

A different method of embryonic nutrition used by some mothers is to feed their developing young via a placental attachment. Only around 18 percent of sharks nourish their embryos in this way, mostly members of the order Carcharhiniformes (ground sharks), such as the Bull Shark, weasel sharks (family Hemigaleidae), and the hammerheads (family Sphyrnidae), but this type of viviparity can also be found in the order Orectolobiformes (carpet sharks). It's thought this method evolved independently as many as 20 times across the different lineages of the shark and ray family tree. The process differs from that seen in humans in that the developing offspring are not connected to the mother's bloodstream via an umbilical cord and placenta. Instead, after the yolk is depleted, its sac develops into a yolk stalk that connects to the wall of the uterus. This yolk-sac placenta transfers nutrient- and oxygen-rich uterine secretions from the mother to the embryos. In some species each shark will grow in a separate crypt within the uterus to stop their umbilical connections becoming tangled. These umbilical stalks attach to the pup's belly between the pectoral fins, and when they are born these little sharks have a small umbilical scar. However, unlike our belly buttons that we keep throughout our lives, in sharks this mark is visible for only a few months before it heals completely.

As they require so much space in their body cavity to house growing young, viviparous female sharks tend to be noticeably larger than their male counterparts, whereas this sexual dimorphism is less pronounced in oviparous and yolk-sac viviparous species. For instance, female Sandbar Sharks (*Carcharhinus plumbeus*) are strikingly larger, reaching 6.6–8.2 ft (2–2.5 m) compared to 5.9 ft (1.8 m) in males. Even though Sandbar Sharks all reproduce by placental viviparity, other aspects of their reproductive biology can vary with geographic location—for example, age of maturity. The gestation period averages from 8–12 months and litter sizes can range from 6 to 13 pups that are 16–30 in (40–75 cm) long at birth.

Viviparity is very costly to the mother both in terms of time and energy, because she must carry her pups inside of her and nourish them as they grow, and pregnancies can also be long. Once again, it varies greatly among species. Some live-bearing sharks, like Sand Tigers, have continuous reproductive cycles, so they can breed every year. In fact, female Sand Tigers spend almost all their adult lives pregnant. Conversely, some species have punctuated breeding; the female will require a long recovery period after giving birth before she can breed again. For instance, female Sandbar Sharks take a year or two off between pregnancies and female Dusky Sharks (*Carcharhinus obscurus*) can breed only once every three years. On the other hand, viviparity means that the pups are born especially big and strong, so they have a very high chance of surviving to adulthood. What's more, this also allows for really large litters in some species. For example, female Blue Sharks (*Prionace glauca*) can have as many as 135 pups per pregnancy.

6

THE WONDER
YEARS

STARTING OUT

It can be hard to imagine that such awesome, fearsome predators as sharks start life as small and vulnerable youngsters, just like us. With no parental care to speak of, newborn sharks (neonates) are born into a terrifying watery world, where they must find their own food and shelter, and somehow manage to evade a plethora of unfamiliar larger predators, from the very instant they first enter this life.

We may be biased, but there is nothing cuter than a newborn shark. Many shark neonates could even be considered doe-eyed, a word most would never use to describe a fully grown shark. They are born as miniature, scale-model replicas of the adults, albeit often with big heads and painfully skinny bodies that must substantially fill out as they grow in length if they are to fulfill their role as a top predator. But with this adorable visage comes an inherent vulnerability that means their entire early life is an independent journey of finding food and avoiding becoming prey themselves. How they go about surviving this battle to reach to adulthood once again varies from shark to shark.

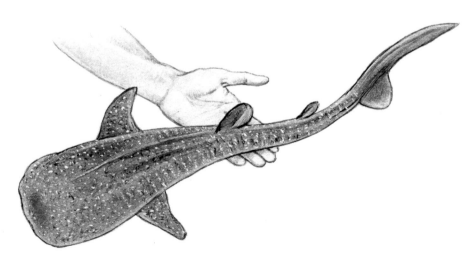

NEONATE WHALE SHARK
Newborn whale sharks are as darling as can be. Despite adults maxing out at 61 ft (18 m) long, they are only about 30 in (75 cm) at birth.

Fake it 'til you make it

Some species of sharks are relatively large when they are born. For instance, Short-fin Makos (*Isurus oxyrinchus*) are about 2 ft (60 cm) long. White Sharks (*Carcharodon carcharias*) and Sand Tigers (*C. taurus*), which are both at least 3.3 ft (1 m) at birth, are already formidable predators in their own right. They must still escape larger beasts, but are pretty good at looking after themselves from the get-go.

A few species of sharks have even evolved defensive weapons to protect themselves. For members of the family Heterodontidae, ten species commonly called the bullhead or horn sharks, newborns wield stout spines on the leading edge of both dorsal fins. Evidence is lacking that the mere presence of these spines deters predators, as sometimes the baby shark, born at an adorable 6 in (15 cm), is swallowed whole by a larger predator, such as a Northern Elephant Seal, Bald Eagle, or larger fish. However, some of these predators will certainly live to regret doing so. One older documentary showcases a Pacific Angel Shark (*Squatina californica*)—a formidable lie-and-wait predator—explosively expanding its maw to suck up an unsuspecting Horn Shark (*Heterodontus francisci*). While viewers bemoan the apparently cruel fate inflicted on the formerly blissfully swimming baby shark, a few tense seconds pass before the Angel Shark violently convulses several times and ejects the lucky Horn Shark, which shakes itself off and continues its journey. Thick skin and stout spines, at least this time, won the day.

For other small sharks, their unimposing physical presence comes with a compensatory behavioral trait—an almost cartoonish, defiant attitude that belies their size. Similar arrogance can be seen in small, nippy dog breeds that growl, "Lemme at 'em!". Atlantic Sharpnose Sharks (*Rhizoprionodon terraenovae*) come into this world with just such an attitude. Despite an unintimidating length of 11 in (28 cm) at birth, neonate and young-of-the-year (that is, in their first year of life) Atlantic Sharpnose Sharks have been known to bite hooks on experimental longlines with fist-sized pieces of bait intended for much, much larger sharks. In a few instances, we have even removed the oversized hook and returned the undersized shark to the water, only to discover the same unabashed youngster on another hook farther down the same longline. Apparently, the initial experience was not harsh enough to be a worthy deterrent. One neonate

NEONATE ATLANTIC SHARPNOSE SHARK
*Despite a diminutive 11 in (28 cm) at birth, young Atlantic
Sharpnose Sharks have a very plucky, defiant attitude.*

Atlantic Sharpnose even tore into a small mesh chum bag containing macerat-
ed bait to attract sharks to the hooks, and contorted itself entirely inside the
sack, like a kid in a candy shop. Neonate Lemon Sharks (*Negaprion brevirostris*)
are also known for their precocious aggressiveness. Given their impressive dentition and innate flexibility, which allows them to bite objects near their tail,
Lemon Sharks are handled with particular care by shark biologists, many of
whom have scars as a result of interactions with juveniles.

The bravado of these and other young sharks has an evolutionary advantage—acting larger than you are to convince bigger fish that it is not worth
messing with you. Not all sharks, however, punch above their weight. Benthic
species like Nurse Sharks (*Ginglymostoma cirratum*) and Smallspotted Catsharks
(*Scyliorhinus canicula*) take advantage of the safety of their bottom environments
and are more secretive, wedging themselves under rock crevices or coral heads
to hide. Some juvenile sharks—for example, nurse sharks—have armorlike
denticles (scales) that also likely deter predators.

Hiding in plain sight

For other young sharks the key to surviving to adulthood lies in hiding in plain
sight. As they are generally less than 14 in (35 cm) long when they hatch out of

their egg cases, juvenile Zebra Sharks (*Stegostoma tigrinum*) have developed a remarkable trick for avoiding their many predators: they mimic venomous sea snakes. By sporting pale, vertical stripes, which break up their black saddles in a distinctive zebra-striped pattern (inspiring their common name), they look a bit like venomous snakes, so predators steer clear of them. This is known as Batesian mimicry, which you also find in butterflies that have evolved markings resembling the eyes of a larger animal, or in flies that have evolved to have distinctive yellow and black stripes like a wasp. It is only when juvenile Zebra Sharks reach 20–35 in (50–90 cm) total length that the dark patches begin to break up into spots, revealing more yellow pigmentation in a spotty-striped pattern. With advancement into maturity, they change again and have a yellow background with dark spots, like a leopard. In fact, these sharks change colors so significantly as they age that scientists originally thought they were several separate species, before DNA analysis revealed the truth.

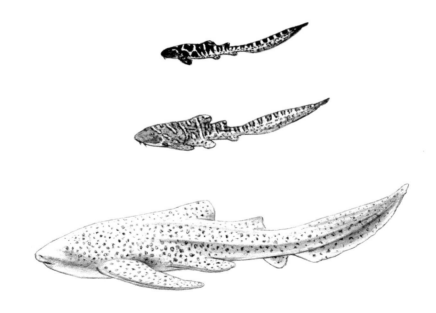

CHANGING COLORS AND PATTERNS OF THE ZEBRA SHARK
*Zebra Sharks undergo such a remarkable change as they age that scientists
once believed newborns (top), subadults (middle), and mature adults
(bottom) to be different species.*

A WATER BIRTH

Although there is no parental care among any sharks, batoids, and chimaeras, this doesn't mean the females simply abandon their young and hope for the best. On the contrary, millions of years of natural selection have ensured that these mothers give birth or deposit their eggs in ways that maximize the survival of their vulnerable young. After all, by this time the females have devoted significant resources to their future offspring, so why not provide them with the benefit of a head start when they first see their new habitat?

Tantalizingly, even the general location of where sharks give birth remains a mystery for the vast majority of species. Unsurprisingly, it is easier to identify these sites for oviparous sharks, as finding their eggs or empty egg cases is a dead giveaway. For example, we know that female Redspotted Catsharks (*Schroederichthys chilensis*) prefer to lay their eggs in tall, structurally complex kelp forests, where they are hidden from predators. However, it is incredibly rare to witness a birth in those species that give birth to live young. A relatively new technique, at least as applied to sharks, employs satellite tags implanted in the uterus via a special applicator. These vaginal implant transmitters (VITs), more commonly called birth-tags, remain in the uterus during pregnancy and are shed during parturition. At that point, they float to the surface and upload location data to a satellite.

Nursery school

Generally speaking, many sharks organize themselves according to size, with younger, smaller sharks living in one habitat and more mature animals occupying a completely different space, with little to no overlap between the two. This is known as size segregation. The regions where the smallest sharks spend their earliest years after birth are known as nursery habitats. These aquatic maternity wards offer the juvenile sharks ample food and protection from predators (including larger members of their own species), and provide the right range of environmental conditions (such as salinity, temperature, oxygen level, turbidity, current, and so on) appropriate for that particular species. Nurseries may also limit competition between sharks of different age classes. As predation risk and competition are often higher in deeper habitats, many shark nurseries are found in shallow ecosystems, such as nearshore bays, mangrove-fringed lagoons, or estuaries. Shallower waters are also often warmer and more productive than adjacent deep waters, which facilitates faster growth of the young sharks.

Given the numerous advantages of nurseries to sharks, it is likely that the use of such areas appeared early on in their evolution. Scientists studying fossilized Megalodon (*Otodus megalodon*) teeth from deposits from 3.6–16 million years ago in the Pacific discerned that their samples were so small (relatively) they must have come from younger sharks only around 8.5 ft (2.6 m) in total length. This suggests that the Pacific Ocean, Caribbean Sea, and the Atlantic Ocean served as nurseries for this mighty, ancient shark.

Determining whether an area constitutes a nursery ground for a particular species is anything but simple. The presence of neonates in an area in accordance with the following three criteria is the gold standard for identifying a nursery habitat: the neonates should appear in greater abundance, enjoy extended stays, and be present annually. For an area to be classified as a nursery ground, it must be used annually, a concept that is sometimes called seasonal site fidelity or philopatry. This means that young sharks can be found there year after year. Additionally, the presence of full-term pups in a pregnant female, detected by dissection or ultrasound, is considered a valid proxy for the location of parturition (birth) sites, in cases where the actual location is unknown. Scientists further classify nurseries as either primary or secondary. Primary nurseries are where birth occurs and the neonates remain for at least a short period of time. In secondary nursery habitats, juveniles remain for a much more extended period—sometimes several years—until they have developed sufficiently to make it in the big wide world.

Yet, while mortality rates might be lower within nursery habitats, this does not mean that all, or even the majority, of little sharks will make it. Within the nursery in Terra Ceia Bay, Florida, juvenile Blacktip Sharks (*Carcharhinus limbatus*) suffer natural mortality rates of 32 to 70 percent in the first 15 weeks of summer and an estimated 35 to 62 percent of neonate Lemon Sharks in their Bahamas nursery will die in their first year. Only the very strongest will survive the twin threats of predation and starvation.

Numerous shark species have been shown to use nurseries. Newborn Black-tip Sharks, for example, frequently aggregate in specific areas in a small bay near Tampa, Florida, during daylight hours, before dispersing at night. In the

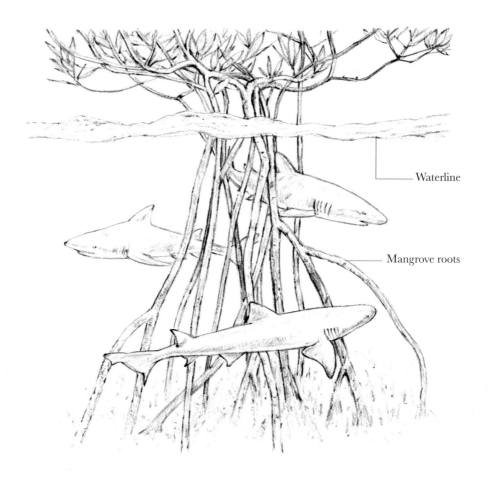

Waterline

Mangrove roots

MANGROVE NURSERY HABITAT
No sharks are cared for by their parents, meaning that many species, including the Lemon Shark, spend their early years sheltered in nursery habitats.

Bahamas and Brazil, juvenile Lemon Sharks aggregate at high tide in their mangrove-lined ecosystem. Sneaking within the tree roots that dip into the water allows the little sharks to avoid patrolling predators that have gained access to habitats closer to shore as the tide rises.

In some species of sharks, the young males and females will grow up in slightly different habitats. For instance, young male (around 70 cm/28 in) Blue Sharks (*Prionace glauca*) aggregate at oceanic seamounts, whereas the females remain in coastal habitats while they mature, for reasons that are not clear, but which may include females foraging for more calorically dense prey found in coastal regions. This might enable them to devote resources to their reproductive system, and also achieve a larger body size with more stored nutrients that would benefit them at maturity.

Among the most biodiverse group of sharks by habitat, the deep-sea species remain the most data deficient in general, and nursery habitats have been identified for only a few species found in the shallowest regions of the deep-sea zone. However, the three criteria cited above were not rigorously applied in designating potential nurseries for deep-sea sharks, a lapse that makes sense when sampling in the deep ocean, but one which raises doubts about the validity of the assertion in some cases. A large quantity of egg cases from an unidentified catshark found in the Gulf of Mexico suggests a nursery may exist at depths of some 1,750 ft (533 m). Similarly, the discovery of abundant catshark egg capsules, most notably those of the Blackmouth Catshark (*Galeus melastomus*), suggests there is a nursery in the Mediterranean at 1,640–2,300 ft (500–700 m). Based on trawl capture of juvenile sharks from similar depths at other locations in the Mediterranean, it has been suggested that this nursery is host to neonate Smallspotted Catsharks, Kitefin Sharks (*Dalatias licha*), and the Velvet Belly Dogfish (*Etmopterus spinax*).

Our protagonists

Among our four main species, the private life of the Sandbar Shark (*Carcharhinus plumbeus*) is the most understood. Estuaries are critical nursery habitats for the growth and development of juvenile Sandbar Sharks (and numerous other species), reducing predation by larger sharks and providing abundant food resources, such as small fish, crustaceans, stingrays, and neonate Atlantic Sharpnose Sharks. Although the largest summer nursery for Sandbar Sharks is found in Chesapeake Bay (in the United States), there are other nurseries on the United States East Coast, from New Jersey to at least South Carolina, as well as similar environments elsewhere in the world, such as Turkey's Gulf of Gökova in the eastern Mediterranean Sea.

The locations of birthing grounds and potential nursery areas for the Spiny Dogfish (*Squalus acanthias*) are not well-known. In the northwestern North Atlantic, they are thought to deliver their six or seven, 8-in (20-cm) pups offshore over a large area during winter. One study collected 47,000 neonates in a small area off southern New England, but otherwise delineation is lacking. Females with full-term pups have also been found in the more southern parts

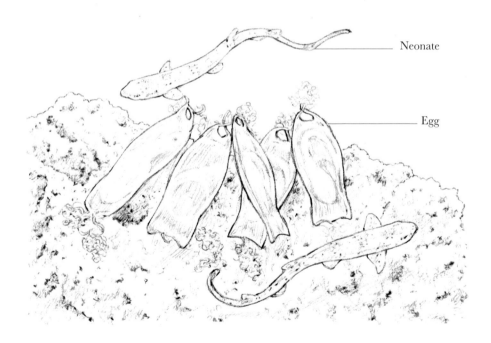

Neonate

Egg

SMALLSPOTTED CATSHARK NURSERY
Female Smallspotted Catsharks lay their eggs in nursery
habitats year-round.

of their distribution in the winter and early spring, implying that birth could occur there as well.

Smallspotted Catsharks are found in the northeast Atlantic and western Mediterranean. The sole oviparous species among our featured sharks, they deposit their egg cases year-round. The scientific literature for this species contains accounts of egg-bearing females and the presence of juveniles in many areas, but nurseries for Smallspotted Catsharks that meet the most rigorous criteria described above have been identified in only a few areas, one being Sardinia, Italy. Since the species is considered by the IUCN as Least Concern, identifying and delineating nurseries may not be a top priority for researchers.

Sitting in the front window of Coastal Carolina University's Boat Operations Center in Georgetown, South Carolina, is a glass display case containing a preserved White Shark nicknamed Baby Jaws, measuring 4.2 ft (1.3 m) and weighing 29 lb (13.2 kg). Since White Sharks are born as small as 3.5 ft (1.1 m), this specimen, which was caught by a commercial fisher off Charleston, South Carolina, in the early 1990s, was almost certainly a neonate and the bait it took may have been one of its first meals. At the time, it was the second smallest documented White Shark.

Since then, as conservation interest in the species has grown, a number of areas of interest have been identified as potential nursery grounds. Catch records of small individual White Sharks from commercial and recreational fishers in southern California, some with yolk-sac scars, suggest that this area might be a nursery for the species, along with the west coast of the central Baja California Peninsula (Mexico), the New York Bight on the East Coast of the United States, the Sicilian Channel, the Aegean Sea, and sites in Australia and South Africa. A 2020 study reidentified a paleo-nursery area for the species along the west coast of South America dating back perhaps 2.5 to 5 million years. The expanded use of drones in the 2020s has resulted in the observation that young-of-the-year and older juvenile White Sharks use nearshore beach habitats in southern California more extensively than previously suspected, in some cases penetrating to within spitting distance of terra firma.

A VIRTUAL CRUISE

Scientists and students have spent over two decades studying the sharks of Winyah Bay, an estuary measuring 25 square miles (65 square kilometers) in the southeast United States. Accompany us on a short virtual cruise to learn what we discovered and how it was done.

We depart from the dock in Georgetown (South Carolina) aboard the 29-ft (8.8-m) research vessel *Eric Thor Koepfler* in the foggy mist of early morning. After reaching our destination—the mouth of the salt marsh-lined Mosquito Creek where it empties into the bay—we set four longlines from the boat's bow, each with 25 hooks baited with fist-sized chunks of Boston Mackerel. Typically, we set our longlines on the sandy or muddy bottom at depths of 6–40 ft (1.8–12.2 m); today, they are at 18 ft (5.5 m).

As Bald Eagles waft on the updrafts above our boat, our 100 hooks, which have fished for only about 45 minutes, yield nine juvenile Sandbar Sharks and a plus-sized Lemon Shark. This is surprising as, in general, the longlines almost always catch one or more of a mix of over ten different species of sharks, from the diminutive neonate Atlantic Sharpnose to 10.5-ft (3.2-m) adult Lemon Sharks. We expediently measure, tag, sample tissues, and safely release the sharks. We return to the dock exhausted but exhilarated by the beauty of what we had just beheld: a wondrous ecosystem and its shark denizens.

Through these methods we have learned something particularly interesting. In the shallower middle portion of the bay, where the water is brackish due to mixing of the fresh water from nearby rivers and saltwater from the sea, our dominant catch is juvenile Sandbar Sharks, and rarely anything else. Moreover, the Sandbar Sharks are almost all four to six years old and measure 2.6–3.1 ft (81–95 cm). We had discovered a secondary nursery for a specific age group of this species that is just now on a trajectory to recovery. There, food is abundant and predators are not. Our studies also showed that the juvenile Sandbar Sharks have an adaptation most of their adult shark predators lack: the ability to tolerate brackish water.

The vulnerable young Sandbar Sharks are not completely protected from predation by the larger sharks in this system. Some Lemon, Bull, Finetooth, and perhaps other species can penetrate the nursery area for short periods of time, but the low salinity is a barrier to them staying in the area for very long. Thus, some young Sandbar Sharks will inevitably succumb to being prey, but that is part of the evolutionary rhythm of life and the continuing saga of the predator–prey arms race.

FRESH FACES

Our understanding of the behavior of juvenile sharks—indeed of their entire biology—is still rudimentary, and we are only just beginning to learn how their sexual, agonistic (competitive), social, foraging, cognition (learning), predator avoidance, navigation, orientation, and communication behaviors function. This is because the logistics of capturing, transporting, and maintaining sharks in captivity (especially larger and/or pelagic species) has made these kinds of studies very difficult to conduct. Yet observing these complex behaviors in the wild is equally limited, both because of the challenging environment (murkiness, depth, temperature, distance, and so on) as well as the effect that the presence of diving scientists has on the species being studied. Therefore, what scientists do know about juvenile sharks is usually for species found in easily accessible habitats or those which can be maintained successfully in captivity. Since every species of shark has a juvenile period, this area remains ripe for scientific study. Additionally, since these life stages are often most susceptible to negative human impacts, the importance of studying them, and protecting both the juvenile sharks and the ecosystems in which they live, is of paramount importance.

Don't I know you?
The successful studies where scientists have overcome these challenges have taught us that young sharks exhibit complex social behaviors. Researchers in

SAFETY IN NUMBERS
*Juvenile Smallspotted Catsharks group together and refuge in their
hidey-holes to avoid larger predators.*

the United Kingdom studying Smallspotted Catsharks in captivity have discovered that juveniles prefer to refuge with familiar individuals rather than strangers. As a result, groups that nestle together to rest in caves or under ledges form based on these predilections. Almost like friendships.

Similarly, captive juvenile Lemon Sharks have been shown to exhibit personal preferences for whom they spend time with. A two-year study of the grouping behavior of 38 juvenile Lemon Sharks in the Bahamas showed the youngsters prefer to socialize with sharks of a similar size that they have already met before, and this facilitates the formation of stable, cohesive groups. Experts suspect that these gatherings protect the juveniles from patrolling predators, most notably larger Lemon Sharks, which can gain access to nearshore nursery habitats as the tide rises. Belonging to a gang means more pairs of eyes (and other senses) looking out for potential predators, decreasing the probability of being the one to be eaten.

The cradle will fall

Nurseries have worked well for numerous species of sharks over time, but aggregations of a large number of juveniles in a defined area can create the potential for disaster if there is a large-scale impact on the system. Consider Lemon Sharks in their nursery ground in the Bahamas. Bimini is one of the most important nurseries in the region for this species and the area also serves to recruit adult Lemon Sharks to Southeastern United States habitats. In the early 2000s, development commenced on a large resort that included dredging and the removal of the mangroves fringing the juvenile Lemon Shark habitat. By 2010, about 166 acres (67 hectares), representing 39 percent of the mangrove habitat surrounding the system, had been removed. Following the early stages of development, the survival and growth rates of juvenile Lemon Sharks decreased, and sharks remaining in the area were less healthy than comparable sharks in undisturbed areas. Reckless development of the small island continues.

Sharks such as the Sandbar Shark that use salt marsh-lined estuaries as nurseries face problems as well. Like mangrove forests, these ecosystems are degraded by a suite of physical human activities, such as pollution, dredging, and the construction of docks, bulkheads, and dikes. Over the past several years, there are signs that the salinity regime of the Winyah Bay Sandbar Shark nursery we discussed on page 128 is changing. Increased rainfall is causing more frequent discharges of fresh water from the rivers that feed the system—a phenomenon which may be associated with climate change. How does this affect the Sandbar Sharks? Studies are ongoing, but lowering of the salinity in the area, even temporarily, may change the structure of the biological community, leading to potential food chain disruptions, as living conditions for the forage base of the system—the crustaceans, worms, small fishes, and so on—begin to shift. Depressed salinities may even lead to the Sandbar Sharks dispersing and the nursery habitat disappearing completely.

At the other end of the scale, climate change is causing some nurseries to move or even grow. For instance, in Pamlico Sound, North Carolina, rising temperatures and salinities mean that Bull Shark (*Carcharhinus leucas*) nursery habitats are actually expanding. Whether this will lead to increased recruitment to the adult population and the extent to which this shift might impact upon the native ecosystem remains to be seen.

IT'S TIME TO GO

Although they are confined to their nursery habitats for their own safety, this doesn't mean little sharks can't move about. Sometimes they might travel quite extensively between different nurseries. How far juvenile sharks roam varies depending on their age, size, species, and habitat. On the lower end of the scale are juvenile Atlantic Sharpnose Sharks, with home ranges as small as 0.5 square miles (1.3 square kilometers) in some studies. At the other end of the spectrum, juvenile Sandbar Sharks in Chesapeake Bay are essentially nomadic, ranging widely each day. Juveniles of oceanic species, such as Blue Sharks and Shortfin Makos, extend their home ranges to include vertical excursions—staying near to the surface in the day and moving slightly deeper during the evening.

As the juveniles grow, their home range typically expands, due to different food requirements and the reduced threat of mortality that accompanies size increases. For instance, in their Bimini nursery habitat, 2.4-ft (70-cm) juvenile Lemon Sharks establish small, well-defined home ranges averaging 0.25 square miles (0.68 square kilometers), but this stamping ground expands to 15 square miles (40 square kilometers) when they reach 6.8 ft (2.1 m) in length.

Making a move
While inshore habitats in tropical and subtropical regions, such as the Bahamas, may be occupied year-round by species like juvenile Lemon Sharks, the

nurseries of many species are not hospitable throughout the year, so the little sharks must venture out on a migration. For example, juvenile Sandbar Sharks inhabiting estuaries north of North Carolina can tolerate temperatures down to about 59°F (15°C), but Chesapeake Bay drops to 41°F (5°C) in the winter, so migrate they must. They spend their winter on the inner to middle continental shelf waters off southern and central North Carolina.

Similarly, those juvenile Sandbar Sharks in Winyah Bay are climatic migrators, leaving the system seasonally, around November, when the water becomes too cold. Some migrate to deeper continental shelf waters off South Carolina. Others mirror the pilgrimages of their counterparts from the northeast United States—following the tourist trail to warm and sunny Florida, where they will remain until April, at which time they (and many people) will move back to their summer homes.

On the return part of their journey, Sandbar Shark migrations are driven by the need to refuge from predators. The climatic conditions in their overwintering location might be suitable for year-round habitation, but the threat of predation if they remain looms large.

Movements of these northern groups of juvenile sharks means that there are two disparate migration routes along the North American East Coast. Since the temperature of the water in North Carolina waters, where the more northern juvenile Sandbar Sharks overwinter, are colder than those of Winyah Bay in winter, from which the southern juvenile Sandbar Sharks migrate, scientists suspect that the two subpopulations might actually have physiological differences in their tolerance to cold.

Coming of age

At what point sharks reach a critical age and size to be able defend themselves and graduate from their nursery once again varies between species. Young Scalloped Hammerheads (*Sphyrna lewini*) stick around their nurseries for a year, whereas Lemon Sharks may linger for two, but some species will stay for several years before heading out into the world. Juvenile Bull Sharks, for instance, might remain in their freshwater nurseries of Florida's Indian River for up to nine years before they completely transition into their offshore habitats. But we've all got to grow up some time. At this point, these little sharks' lives change forever, as they must enter their adult world.

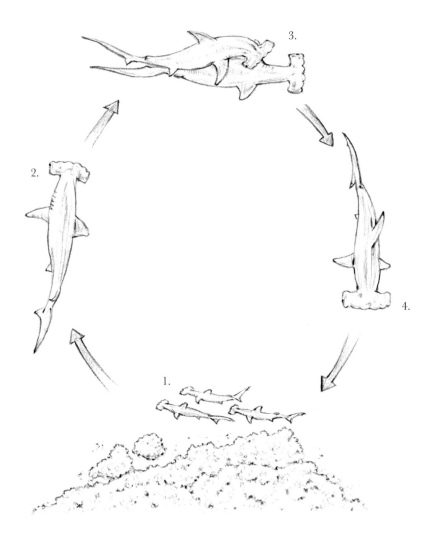

THE LIFE CYCLE OF SCALLOPED HAMMERHEADS
Young Scalloped Hammerheads grow up in coastal nursery habitats (1) before making their way offshore as subadults (2). When they reach maturity and mate (3), the females will return to their nursery habitat to give birth to their own young (4).

7

THE ADULT
YEARS

THE TERRIBLE TEENS

If you're a juvenile shark and natural selection has been kind to you (and you've had a bit of luck on your side), by the time you reach adulthood you will have already survived a veritable minefield of dangers. Beginning in the womb, you could have been eaten by a more mature sibling, in the case of Sand Tigers (*Carcharias taurus*), or outcompeted for a sumptuous meal of unfertilized ova, if you are a Shortfin Mako (*Isurus oxyrinchus*). For placental viviparous species, such as the Sandbar Shark (*Carcharhinus plumbeus*), there is always the risk of being strangled by the life-giving umbilicus: hoisted by your own petard. If you were to develop in an egg laid by your oviparous mother, you could be feasted upon by an ocean of predators, or even swept away by waves or currents to less clement environs. When you do finally burst onto the scene as a neonate, a new suite of threats awaits, mostly in the form of predators, but also an array of abandoned and active fishing gear. As a juvenile you must bravely face heading away from the safety of your nursery habitat into a whole new world.

Welcome to adulthood—almost. At this point, sharks, which are now called subadults, may be sufficiently grown to strike out into the world, but they are not yet sexually mature and able to reproduce. They must continue to forage and grow until they are able to have young sharks of their own.

Grow a spine

In contrast to what we see in many animals, sharks do not achieve a final adult size at which point their growth halts. In fact, sharks grow continuously throughout their lives, a phenomenon called indeterminate growth. Therefore, a larger shark is typically an older shark. That said, each species does seem to have a maximum size and age where growth becomes so imperceptible that it may just as well have stopped. Spiny Dogfish (*Squalus acanthias*) max out at around 4 ft (1.2 m) for females and 3.3 ft (1 m) in males. And despite what you have seen in some horror movies, the existence of White Sharks (*Carcharodon carcharias*) larger than 21 ft (6.4 m) has never been validated.

That raises a question: How do we know how old a shark is? Sharks in aquaria have provided invaluable data in this area for a handful of the species that can be maintained in captivity, but extrapolating these data to sharks in the wild, as for most zoo animals, is not valid, as life in captivity changes everything, biologically speaking. Recapturing tagged sharks in the wild has also yielded good data on growth, but recapture rates are in the low single percentages and this only declines the longer a tagged shark has been at large. For the most reliable way to age sharks, scientists have turned to trees and essentially use the same method as botanists: ring counting. As sharks grow, they deposit additional layers around their vertebrae, similar to the growth rings of, say, a Sequoia. Where some sharks possess other hardened parts, like the fin spines of the Spiny Dogfish, ecologists can use these instead.

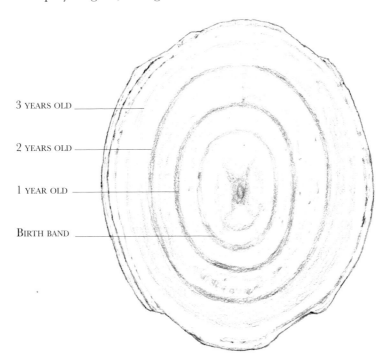

3 YEARS OLD

2 YEARS OLD

1 YEAR OLD

BIRTH BAND

VERTEBRAL GROWTH BANDS
*Sharks lay down new rings of growth over their existing skeletons
in order to grow larger, which can be seen as bands in their
vertebrae and other hardened parts.*

137

Through ring counting, scientists have discovered that growth rates can vary massively between species, but most are consistent with the conservative, or slow, life history strategy of most of the shark world. While many deposit a new ring of growth every year—Sandbar Sharks, for instance—other species might put down a new ring only every two years. Greenland Sharks (*Somniosus microcephalus*) grow at a glacial rate of only 0.25–0.5 in (0.5–1 cm) per year. Atlantic Sharpnose Sharks (*Rhizoprionodon terraenovae*), at the other extreme, grow so fast you can almost see it happening before your eyes, with pups increasing in size by as much as 2 in (5 cm) a month during their first three months of life. This allows them to reach sexual maturity at a mere three years of age. But this is the exception among sharks.

The Sandbar Shark, one of our featured sharks, exhibits the conservative life history characteristics typical of sharks. It takes females about 15 years to grow from 16–20 in (45–50 cm) at birth to sexual maturity at approximately 4.2–6.2 ft (1.3–1.9 m), an average growth rate of about 2.4 in (6 cm) per year. Males mature at 3.9–5.9 ft (1.2–1.8 m). Maximum size is about 8 ft (2.4 m) and they live 30–35 years.

You would think that the life history characteristics of a species as iconic as the White Shark would be well-known, but, in fact, there are still some gaps in our knowledge. We do know that White Sharks are born at 3.6–5.2 ft (1.1–1.6 m) and grow to about 20 ft (6 m) and 4,200 lb (1,900 kg) over their lives, but we are not completely sure how quickly they do so. We also don't know for sure how long they live. Estimates range from just 33 to over 70 years.

Growing up

Even within one population, sharks at different life stages grow at varying rates, and it is also quite common for males and females to grow at contrasting speeds, to reach maturity at different sizes. For instance, Spiny Dogfish males mature earlier (at around six to ten years of age) than the females (12–16 years). Female Sandbar Sharks, Spinner Sharks (*Carcharhinus brevipinna*), and Dusky Sharks (*Carcharhinus obscurus*) grow much more rapidly than the males, and they all grow the most quickly in the first three years after birth. Scientists suspect this allows the young sharks to reach a less vulnerable size as soon as possible and to maximize their reproductive output.

It is also quite common for different geographic populations of the same shark species to grow at completely dissimilar rates and reach sexual maturity at different sizes (that is, ages). Smallspotted Catsharks (*Scyliorhinus canicula*) in the Mediterranean Sea are smaller than those in the North Atlantic and the North Sea, for example. These differential growth rates are due at least in part to differences in habitats; water temperature, prey availability, predators, and even the density of conspecifics (sharks of the same species) can all vary and affect how quickly sharks grow. There may also be a genetic basis for the different rates, since populations may be genetically distinct. One implication of this knowledge is that scientists can't assume sharks from one part of the world will be able to withstand exploitation at the same rate as their counterparts across the globe, or maybe even those more closer by.

Understanding growth rates and longevity is critical to protecting threatened sharks because it gives us some idea how quickly the species may be able to bounce back after serious population declines and, therefore, how strict conservation measures need to be. For instance, Blue Sharks (*Prionace glauca*) are exceptionally fast-growing (for sharks, that is) and they become sexually mature when only 6.6 ft (2 m) in length, at just five to six years of age. They may live for an estimated 26 years. Comparatively, the Narrownose Smoothhound (*Mustelus schmitti*) grows more slowly, reaches sexual maturity later in life (six to seven years), and dies younger (at around 21 years). This equates to a reduced reproductive potential for the Smoothhound (fewer offspring over their life span) and a depressed rebound potential (the ability to recover after stock reductions or collapses) compared to the Blue Shark.

ON THE MENU

While sharks are all predators, not all are apex predators. In fact, the majority of shark species are not apex predators—another myth exploded. An apex predator is an animal, typically large, at the very top of its food chain, at the highest trophic level, consuming other animals and having very few (if any) natural predators. As adults, the Great Hammerhead (*Sphyrna mokarran*) and Oceanic Whitetip Shark (*Carcharhinus longimanus*) are apex predators (but are mesopredators as juveniles), as is the Oceanic Whitetip Shark (*Carcharhinus longimanus*). In any ecosystem, apex predators are the least numerous animals and thus the most vulnerable to many threats in their system. Their loss can impact the entire community within an ecosystem.

Only a small handful of the 540-plus different species of sharks that we know of today are apex predators. The White Shark is the undisputed apex predator among our four focus species. Its diet includes marine mammals, sharks, and other large fish. Yet there can be, and often is, more than one apex predator in an ecosystem. Off South Africa, White Sharks are not the only apex predators. In parts of their range, they overlap with Broadnose Sevengills (*Notorynchus cepedianus*) and Tiger Sharks (*Galeocerdo cuvier*), two species that also sit atop the food chain, and they all compete for many of the same resources. What's more, the recent immigration of Orcas into their ecosystems means that White Sharks now even have their own predator in South African waters.

Most sharks are lower on the food web, so they are described as mesopredators (meaning "middle predator"). Sandbar Sharks are mesopredators of small bony fishes and occasionally smaller sharks, rays, crustaceans, and cephalopods (squid and octopuses). On coral reefs, most sharks are considered mesopredators. Whereas Bull Sharks (*Carcharhinus leucas*), Silvertip Sharks (*C. albimarginatus*), and Tiger Sharks are apex predators, the medium-sized sharks, such as Caribbean Reef Sharks (*C. perezi*) and Grey Reef Sharks (*C. amblyrhynchos*), and smaller species such as the Blacknose Shark (*C. acronotus*) and Halmahera Epaulette Shark (*Hemiscyllium halmahera*), are all mesopredators.

PYJAMA SHARK CHASING AN OCTOPUS
The vast majority of sharks are not apex predators but are mesopredators
and hunt smaller animals like octopuses, squid, shellfish, and small fishes.

Our Smallspotted Catshark is a mesopredator, preying on fishes, mollusks, crustaceans, squid, octopuses, worms, and snails. So is the Spiny Dogfish, which chows down on shrimp, crabs, squid, and small fishes like herring. The Pyjama Shark (*Poroderma africanum*) might be a voracious predator, renowned for aggressive hunting of octopuses and squid, yet they too are not the apex predator in their habitat, which may include Tiger Sharks, Broadnose Sevengills, and White Sharks.

On the level

Sharks that feed on plankton, like the Basking Shark (*Cetorhinus maximus*), occupy a lower trophic level than even the mesopredatory sharks. This means that, despite being the largest fish in the ocean, filter-feeding Whale Sharks (*Rhincodon typus*) are lower on the food web than the smallest species of shark in the world, the deep-sea Dwarf Lanternshark (*Etmopterus perryi*), which eat shrimp and fishes that are at a higher trophic level than the forage base of the plankton-eating sharks.

Trophic levels are not static, however. The level a species falls into may shift seasonally, and sharks can also completely change trophic levels as they grow. It is also not surprising that juvenile sharks are at a lower trophic level than adults.

141

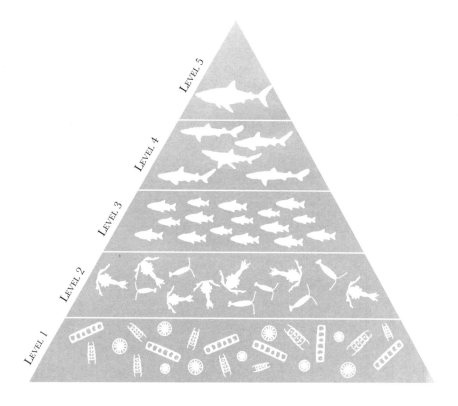

GENERALIZED TROPHIC PYRAMID
*Sharks occupy several different levels on their respective trophic pyramids;
some are apex predators, others are mesopredators that feed on fish and
invertebrates. Plankton-eating sharks are not shown here.*

As young Tiger Sharks grow, they shift from eating small bony fishes and cephalopods and incorporate more turtles, larger bony fishes, other sharks, and even
mammals and birds into their diet—bumping them up the food web.

On a five-point scale (that is, in an ecosystem with five trophic levels; see illus-
tration above), with higher numbers associated with higher-level predators,
trophic levels of mesopredatory sharks range from about 3.1 for the Zebra
Shark to 4.3 for the Caribbean Reef Shark. Trophic levels for sharks considered
apex predators range from 4.4 for the White Shark to 4.7 for the Broadnose
Sevengill Shark. If we consider all species that have been assessed, sharks'
trophic levels are, on average, greater than 4. For comparison, the trophic level
for Orcas is 4.5.

Many sharks, particularly oceanic sharks, which live in a food-poor environment, frequently consume indigestible foreign objects. When a Blue Shark, for instance, encounters tin cans or unopened bottles of wine (both of which have been found in their guts), it must assume that these represent a lucky gastronomic discovery and conclude, "Oh well, better eat them, since I don't know when my next opportunity to feed will come." Tiger Sharks are often called the garbage cans of the seas because they have been known to ingest license plates, cans, plastic trash, tires, burlap sacks, rags, shoes, bottles, coal, nails, video cameras, and, in one incident, even a chicken coop!

Engorging the gut with inedible items is certainly not good for these sharks, as they can cause internal injuries, may be toxic, or could block their digestive tract completely, leading to starvation, a major cause of death for predators. Luckily though, many sharks have an ingenious adaptation that allows them to expel indigestible objects: they can temporarily evert their stomach out of their mouth, rinsing any bones, parasites, mucus, and foreign bodies back out to keep their stomach clean. Caribbean Reef Sharks can perform this maneuver in as little as around 0.3 seconds.

If sharks ingest items that they cannot eliminate in their feces, they may also be able to rinse the other end of their digestive tract: by extruding a part of their intestine out of their cloaca. Scientists have witnessed a wide variety of shark species—including the Blacktip Reef Shark (*Carcharhinus melanopterus*), Lemon Shark (*Negaprion brevirostris*), Bull Sharks, and Sicklefin Lemon Shark (*Negaprion acutidens*)—everting an inch (2.5 cm) or more of their intestine for minutes at a time. Expulsion into the ambient seawater can flush out any trapped detritus, rather like an external colonic irrigation.

All you can eat

How often does a shark need to eat? How much do they eat? Are sharks food generalists or specialists? The answers to these questions depend on the specific shark and its ecological role, life stage and size, reproductive status, activity level, and location. The first is the easiest question to answer. Generally, sharks do not need food as regularly and consistently as humans. In fact, they can go for days, and maybe even months, between meals, surviving on energy stores of fats cached in their livers. Once again, the stats are unique for each species. Scientists have observed a healthy Swell Shark (*Cephaloscyllium ventriosum*) fast for over a year. On the other hand, an enormous filter-feeding Basking Shark may spend many hours of each day feeding.

How much food is needed? On average, sharks consume 2–3 percent of their body weight per day, but again, this varies. It has been estimated that the Greenland Shark—a slow-growing, long-lived, cold-water species with an extremely slow metabolism—requires less than 0.4 lb (0.2 kg) of food daily. At the other extreme, a typical 2,200-lb (1,000-kg) White Shark—a much more active species with an elevated metabolic rate—needs to eat over 40 lb (18 kg) of prey a day. They have been known to devour 66 lb (30 kg) of whale blubber in a single sitting.

Many sharks are generalists and incorporate a wide variety of different prey items into their diet. This flexibility means they can switch prey in ecosystems

SHARKS SCAVENGE ON THE CARCASS OF A WHALE
Many sharks, like these Oceanic Whitetips, Blue Sharks, and Silkies,
will scavenge on dead animals that they come across.

where the types and abundance of prey vary, either naturally or because of human impact. Spiny Dogfish, for example, are opportunists that eat whatever prey happens to be most abundant and available. Many sharks are also scavengers and will feed on the carcasses of other animals they encounter. We've all seen videos of oceanic sharks eating at a whale carcass buffet. Comparatively, some species are ecological specialists, favoring a specific prey item when it's available. Bonnetheads (*Sphyrna tiburo*) love eating crabs. Hammerheads (*Sphyrna* species) forage on stingrays. Their otherworldly cephalofoil (the technical term for their unusual head) acts as a forward rudder, enabling quick turns in pursuit of nimble prey, and they have been observed using their head to pin rays to the substrate as they devour them.

Grabbing a bite

Some benthic sharks, such as the bullhead sharks (family Heterodontidae), Leopard Sharks (*Triakis semifasciata*), and Whitespotted Bamboo Sharks (*Chiloscyllium plagiosum*), have developed a very specialized feeding method, one superficially similar to that of their jawless ancestors: they draw their food into their mouths by creating suction. Nurse Sharks (*Ginglymostoma cirratum*) hunting at night use their sensory barbels to close in on prey and then expand their prodigious buccal (cheek) muscles to suck small mollusks and crustaceans into their mouths, or even to pull a Queen Conch from its shell—rather like a vacuum cleaner. It then proceeds to macerate larger items through a series of bite-and-spit maneuvers that result in sufficiently small, softened morsels suitable for swallowing.

Pacific Angel Sharks (*Squatina californica*) have developed an especially clever feeding strategy. Being nocturnal, they have learned to track the movements of bioluminescent (light-producing) organisms, like dinoflagellates (single-celled aquatic organisms) to find their dinner. As their prey moves through the water, disturbing these tiny organisms and causing them to light up, the shark is able to spot the eddies and flows of the specks of light. They cannot see their prey directly, but they have learned that the movement of the luminescence means that a potential meal is nearby. They use their electrosenses to pinpoint the prey's exact location, and then pounce.

THOSE TEETH AND JAWS...

As we mentioned in Chapter 1 (see page 21), all sharks have polyphyodont dentition, in which multiple rows of teeth in various stages of development can advance like a conveyor belt. The teeth, as well as their arrangement, are as diverse as the sharks themselves. Variations include teeth which may be flattened, needlelike/spindly, triangular, and with or without serrations. You can infer much about the diet of a shark from the morphology of its jaws and teeth.

In numerous species, such as Blue Sharks, the teeth are sharp and serrated, perfect for shearing chunks of flesh from prey—mostly bony fish, squid, and mammal carcasses. The serrated teeth of Tiger Sharks, which eat larger prey, are jagged on both edges and arranged sideways, leaning slightly inward, so that the tips point toward the back of the mouth. As the jaws of Tiger Sharks are surprisingly flexible and bend when contacting prey, they create an almost perfectly uninterrupted cutting edge when the jaws close on their prey.

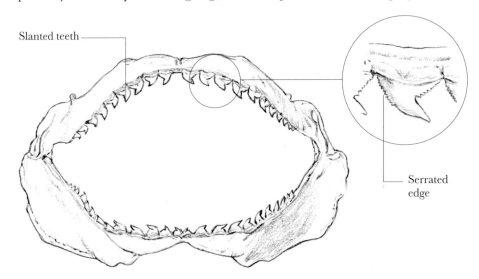

Slanted teeth

Serrated edge

TIGER SHARK DENTITION
The backward-leaning shape of the teeth and the flexibility of their jaws mean that when Tiger Sharks bite down, there is a continuous cutting edge in contact with their prey.

Smallspotted Catsharks have surprising large teeth arranged such that each row is slightly askew of the row behind and in front, like seats in a theater. They forage primarily on smaller bony fish, crustaceans, mollusks, and other marine invertebrates. Other species have very spindly, sharp teeth. Consider the iconic Sand Tiger. Adored in aquaria all around the world because their loose gape and protruding snaggleteeth are fearsome to view, these sharks feast on small fish and are not really a threat to humans. Their rearward-facing teeth are perfect for gripping their slippery prey, which impale themselves evermore as they struggle to escape. Mako sharks (*Isurus* species), wobbegongs (family Orectolobidae), and many deep-sea sharks sport similarly elongated, recurved teeth that function similarly.

Durophagous sharks—that is, those that eat predominantly hard-bodied prey—typically have smaller, platelike teeth adapted for crushing. Houndsharks, like the Starry Smoothhound (*Mustelus asterias*), have broad, flattened teeth, which are arranged in an interlocking pattern, perfect for grinding up the hard parts of their shellfish prey.

Similar to the specialized teeth of humans (incisors, canines, and molars) that all perform their own jobs, numerous species of sharks have different teeth in various regions of their mouths. Many carcharhinid sharks—for example, Sandbar Sharks and Blacktip Sharks (*Carcharhinus limbatus*)—have narrow, cusped lower teeth for grasping prey, whereas the upper teeth are slightly wider, with sharper edges that allow them to slice prey into pieces. This heterodonty (the Greek for "different teeth") is most notable in the bullhead sharks, whose varied dentition earned them their family name Heterodontidae. All ten species have cusped teeth in front for grasping and molariform teeth in the rear for crushing prey like sea urchins.

In White Sharks, the teeth of the upper jaw are wider and more triangular compared to the pointier, narrower teeth in their lower jaw. This is ideal when scavenging on the carcasses of large prey like a whale. When these opportunities arise, White Sharks will wedge their bottom teeth into the flesh, thrust their jaws out, and shake their head, using their saw-like uppers to slice off a semicircle of blubber.

Many sharks also undergo an ontogenetic diet shift as they grow, whereby they completely switch their prey. The Shortfin Mako swallows some of its prey, which include bony fishes and cephalopods, whole. However, as they age, their teeth become broader and flatter, enabling them to widen their prey options to include organisms such as Swordfish, tuna, sharks, sea turtles, and marine mammals. These may be too large to swallow whole, but they can remove chunks of flesh with their cutting teeth.

Lastly, and ghastly, shark biologists are often consulted to determine if a wound, typically on a person's torso or extremity, is in fact a shark bite. Puncture marks in an arc are a giveaway that indeed the bites were caused by a shark, and the size and shape of the arc can also be diagnostic in determining how large the shark was. Since teeth vary among shark species, as well as between the upper and lower jaws of some species, any impressions left by the puncture marks and the spacing between them can also be invaluable. However, even under the best of situations, because there is no single database of shark bite forensic information, identifying the kind of shark responsible is usually an educated guess at best.

In suspense

It is not only their teeth that determine how and what sharks eat. The morphology of their mouth and jaws is also pivotal. Excluding the Whale Shark, Megamouth (*Megachasma pelagios*), and the frilled sharks (family Chlamydoselachidae), most sharks have jaws that could be described as underslung, or subterminal. A major advantage of this type of jaw is that it can accommodate heavier jaw musculature, for greater jaw mobility and protrusibility.

In humans, the upper jaw is fused to the skull, so we can't really project our jaws. Try it for yourself. You can take a bite out of something bigger than your gape, but unless the food is compressible, good luck squeezing the whole shebang into your mouth. If sharks possessed a similar jaw suspension, they would be a minor group—like the lampreys and hagfish—or would perhaps exist only as fossils, as evolutionary experiments gone wrong. However, as their jaws are more loosely connected to the overlying chondrocranium (skull) and supported only at the rear and corners, sharks have a much wider gape, so they can target a broader range of prey items. It is this protrusibility that allows White Sharks to so impressively thrust their jaws around an enormous mouthful of blubber as they scavenge on a whale carcass. The significance of this larger gape for the success of modern sharks cannot be overstated.

Although all sharks can protrude their jaws somewhat, there is variability, with the least protrusible jaws found in the more primitive sharks, including the frilled sharks, cow sharks (family Hexanchidae), and horn sharks (family Heterodontidae), as well as extinct species. Greater protrusibility and jaw mobility are found in the requiem sharks (family Carcharhinidae), such as Sandbar Sharks, whose jaw suspension is among the most mobile of all sharks. The most extreme jaw mobility is found in the batoids. Their jaws are so loosely supported by their skeleton that when we catch a large ray on a longline for our research, we do not hoist it into the boat for fear of dislocating its jaws.

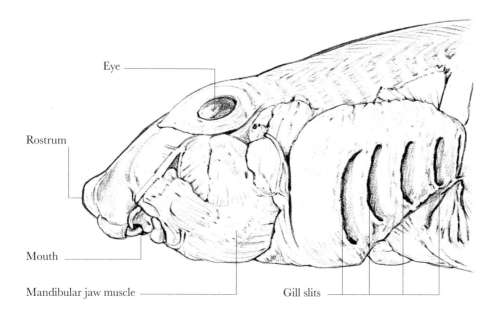

JAW MUSCLES OF A HORN SHARK
The heavy musculature of the powerful jaws of the Horn Shark (shown with the skin removed) provide the force it needs to consume hard-bodied prey.

NOT ALL WHO WANDER ARE LOST

The spatial scales over which different species of sharks roam are as varied as all other aspects of their ecology. As we explained for juvenile sharks, some species have very small home ranges that they barely ever leave (see page 132). Take Smallspotted Catshark, for example. They are nonmigratory and appear to form localized, somewhat genetically distinct populations across their expansive range.

On the other hand, many different species of sharks migrate, including Tiger Sharks, Leopard Sharks, and Smooth Hammerheads (*Sphyrna zygaena*), but each species travels to its specific destination over a different spatial scale. The most migratory species is likely the Blue Shark, which can travel over 5,700 miles (9,000 km) annually. Most famously, a White Shark named after the actress Nicole Kidman made history when she was recorded having swum some 12,000 miles (20,000 km) from South Africa to Australia and back again in just nine months. In fact, she managed the journey one way in a mere 99 days, making for the fastest transoceanic migration ever recorded for a shark.

Undertaking extended migrations is costly in many ways. First, the long journey is time-consuming, so it may take time away from other activities. Second, it may be exhausting. Swimming requires a huge energy investment, which can take a toll on the body, so many sharks will be noticeably skinnier after their travels. Similarly, migrating is risky. Leaving the home range and striking out into distant territory means a shark may have to cross inhospitable regions, go for extended periods without food, or be exposed to unfamiliar, dangerous predators, not to mention fish nets. A relatively recent threat, especially for the plodding and often surface-feeding Whale Sharks and Basking Sharks, is collisions with ships and propellers. Yes, today sharks are becoming roadkill.

Moving on

Given the natural threats that sharks face by migrating, the trip will have been vetted by natural selection and determined to be worth the investment. For many species of sharks, the specific reasons for migration remain unclear. For instance, Basking Sharks equipped with satellite tags in Massachusetts, in the United States, have been tracked swimming all the way across the Atlantic Ocean, about 2,900 miles (4,700 km) away through barren ocean, to the Hebridean Islands in Scotland. Whether they journey to find food or a mate, or maybe both, we don't know for sure.

Generally speaking, migratory animals, including sharks, travel for a reason: namely refuge, food, and/or sex. Sandbar Sharks migrate for all these purposes. Spiny Dogfish are highly migratory along the United States East Coast, moving from their summer residence north of Cape Cod, Massachusetts, to as far south as South Carolina in late fall, to spend the winter in more comfortable temperatures.

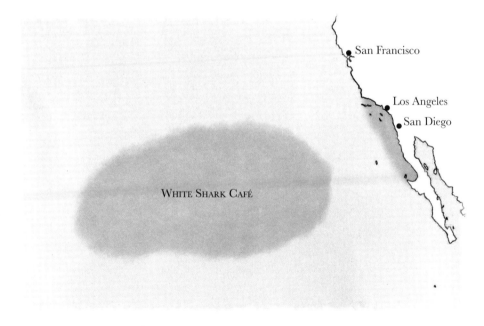

WHITE SHARK CAFÉ
Scientists remain somewhat mystified about why White Sharks from Mexico and California migrate annually to this obscure offshore area in the Pacific.

One of the most fascinating of all shark migrations is that of the White Sharks in the Pacific Ocean. Every summer, sharks from Guadalupe Island, Mexico, and the Farallon Islands, California, undertake an enormous migration, swimming at least 74 miles (119 km) per day, away from the prey-rich coastal regions, to spend several months aggregated at an obscure but expansive offshore location seemingly in the middle of nowhere, known as the White Shark Café, or the Shared Offshore Foraging Area (SOFA). After three to fifteen months hanging out there, both groups turn around to make the return journey, always arriving in their respective coastal homes at around the same time. No sharks go back to the wrong place: a Guadalupe shark always goes back to Mexico and a Farallon shark to the United States. Although there is no definitive explanation for this journey, recent evidence suggests that it may be to forage.

For some sharks, migrations in search of food are not horizontal. Instead, they travel on the vertical plane to track down food. The Megamouth Shark makes a daily journey up and down through the water column (known as diel vertical migration), in pursuit of minute organisms to eat. This daily vertical migration is the largest of all migrations on Earth in terms of scope and the numbers of individuals participating. Every day, throughout the world's oceans, countless microscopic zooplankton swim up to shallower depths at night and return to deeper waters during the day, and predators, like dolphins, seals, bony fishes, and sharks, follow them in droves.

Many sharks migrate for the purposes of reproduction. In some cases, at the end of an extended journey, there is a big aggregation of many sharks—basically a huge party to find a suitable mate. Nurse Sharks are known to travel to very specific locations to mate and will return to the same spot year after year for the same purpose. Blacktip Sharks can be seen in huge numbers off the coast of Florida during the summer months where they get together to … well … get together! They migrate all the way there in order to find a mating partner.

DO YOU WANNA BE IN MY GANG?

While many migratory sharks in search of food or mates temporarily aggregate in large numbers, other species form permanent groups. While there are a myriad of different reasons why each species has evolved to live in a group for adult sharks, they all generally boil down to two broad and very basic categories: food and sex—what other reasons are there to go to a party? For example, group living might provide better foraging opportunities or increase hunting success, or it might mean there are more opportunities to reproduce or a higher chance of surviving to breeding age. For small-bodied mesopredatory species, aggregations may confer another benefit, reducing predation, since there is safety in numbers.

SPINY DOGFISH HUNTING IN A PACK
Spiny Dogfish may hunt in large groups (up to 900 strong), working together to sweep an area for food.

There are many examples of sharks cooperating in order to forage in groups. For instance, Whitetip Reef Sharks (*Triaenodon obesus*) hunt around coral reefs in groups as large as 900. Spiny Dogfish also forage in huge packs. Sometimes hundreds or even thousands strong, they will sweep across an area, eating any and all smaller fishes in their path. For both species, this behavior not only improves their foraging success, but also offers these smaller sharks protection from larger predators.

While they are predominantly solitary animals, White Sharks are also sometimes known to forage in pairs, likely so they can eavesdrop on each other. When one of the pair senses a bleeding or thrashing animal, representing potential prey, the other member of the pair, which has actively stayed nearby, piggybacks on the success of its partner. It is not a coordinated attack, but this strategy doubles their chances of a successful hunt.

Blacktip Reef Sharks in Polynesia have been shown to live in complex groups with extremely fine-scaled social structures. Within their extended community— of over 100 individuals—sharks form consistent subgroups made up of the same individuals that regularly interact with each other and share the same space. Blacktip Reef Sharks learn how to avoid fishers when they have previously been caught, and there is evidence that they will actually communicate this to other members of the group. Therefore, scientists suspect that group living is beneficial because it allows sharks to share social information vital for their survival.

Grey Reef Sharks may also live permanently in large groups. These associations also allow for social communication, with sharks helping each other find the best foraging spots. Around the Palmyra Atoll in the Pacific Ocean, these communities have complex structures, and scientists have learned that the sub-groupings remain consistent and stable over many years, with certain individuals consistently interacting regularly with the same cast of characters within their group. Basically, it seems they form affinities not unlike friendships between individuals.

8

SHARKS
THAT BREAK
THE RULES

BREAKING THE MOLD

The remarkable display of diversity that we see among living animals today is not an accident or the result of some grand design but has arisen thanks to evolution, which can be considered as a series of risk-taking experiments to determine what works and what doesn't in an organism's form (morphology), function (physiology), and/or behavior. A newborn cheetah that shares all the adaptations both it and its litter mates will need as an adult (such as a small skull, flexible spine, long legs, and lightweight skeleton), but born with a small heart, would likely perish since it could never accelerate to the speeds required to catch an impala or springbok.

In general, traits that survive the evolutionary challenge may create an animal that is a superior hunter, can feed in a different way, use a novel defensive strategy, camouflage itself well, communicate with or court mates better than its counterparts, and so on. Over evolutionary time, small changes can build into massive overhauls. As a result, we see some truly bizarre adaptations that may look as if they were dreamed up on the set of a sci-fi movie. Some may seem maladaptive, yet they persist because evolution favors traits that are beneficial on balance—in other words, those that help the organism to survive in some way and so pass its genes onto the next generation. Sharks play by this same set of rules.

We hope you've enjoyed what you have learned so far about sharks, but perhaps this was quite stereotypical and you now see that sharks are more diverse than the media's steady stream of White Sharks (*Carcharodon carcharias*), Tiger Sharks (*Galeocerdo cuvier*), and reef sharks (*Carcharhinus* species) might suggest. So, in this chapter we would like to take the opportunity to introduce you to some of the most spectacular oddball sharks. These are the species where evolution really broke the mold, both in looks and behavior. To some they may seem so unsharklike as to make you question your allegiance to sharks as a group, but we think otherwise, and suspect that these fascinating weirdos might just end up becoming some of your very favorites.

What a pig

Beauty is in the eye of the beholder, so depending on who is looking, you might call the roughsharks (family Oxynotidae) adorably charming or absolutely repulsive. We'll let you guess where we fall! These strange little sharks have beautifully bizarre, sail-shaped dorsal fins and are triangular in cross section when you look at them directly from the front. But most people probably wouldn't notice because they would be too distracted by their blunt face, thick lips, and large-set nostrils, which earned one species the common name of Pigfaced Shark (*Oxynotus centrina*). Poor little guy.

Like the other four described species of roughsharks, Pigfaced Sharks predominantly inhabit the deep sea and their biology is poorly understood. We know they may reach 4.9 ft (1.5 m), but most specimens are smaller than 3.3 ft (1 m) and, like many deep-sea sharks, they have very small gill slits, which reflects the low demand for oxygen associated with their slow lifestyle. Pigfaced Sharks are rarely encountered. In fact, they are so little known that when one was caught in the Mediterranean Sea, many people online claimed it was hoax and the images were faked. They are (thankfully) very real.

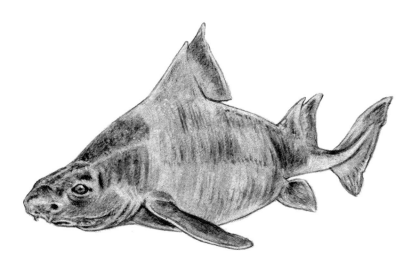

PIGFACED SHARK
*To some, these unusual little sharks are aptly named, possessing blunt,
wide snouts as well as unusual fins.*

157

Seesaw

Some of the weirdest-looking sharks are the sawsharks (order Pristiophori-formes), so called thanks to the huge, sawlike rostrum projecting in front of their faces. This weapon has evolved for hunting and can also be used defensively. For instance, the Longnose Sawshark (*Pristiophorus cirratus*) uses its saw to strike out and immobilize small crustaceans, fish, and squid. Currently ten different species of sawsharks have been described. None are very well-known (especially with regard to their behavior) and some are so recently discovered that we know virtually nothing about their biology. Sawsharks are generally found in shallow waters in temperate and tropical regions (although some tropical species are found deeper), where they sometimes form large schools. The Longnose Sawshark may grow to 4.9 ft (1.5 m), while the African Dwarf Sawshark (*P. nancyae*) maxes out at only 2 ft (60 cm).

Many people confuse sawsharks (true sharks) with sawfishes (batoids that have evolved a sharklike body), as both groups are armed with impressive toothed rostrums, but if you know what to look for, it is very easy to tell them apart. Aside from the gill placement (on the underside in sawfishes and on the sides in sawsharks), sawfishes lack sensory barbels, which hang from the front of all sawsharks' faces and may be used to locate prey by detecting changes in texture of the surface of the seafloor, as well as potentially sensing a prey item's wake. The rostral teeth are also noticeably different between the groups. In sawfishes, all the teeth are the same size (homodont), whereas sawshark rostral teeth alternate between smaller and larger down the length of the rostrum (heterodont). Their teeth in their mouths are also a dead giveaway. Sawsharks have small, spiked teeth, perfect for grabbing onto wriggly fish and squid, whereas the sawfishes' flattened teeth are ideal for grinding up crustaceans and mollusks. However, the most obvious difference between the two groups is their size. Sawsharks are much smaller and generally have a much more slender build compared to their batoid look-alikes. In fact, sawfishes are some of the biggest fish in the oceans, with Green Sawfish (*Pristis zijsron*) reaching a whopping 24 ft (6 m) in length.

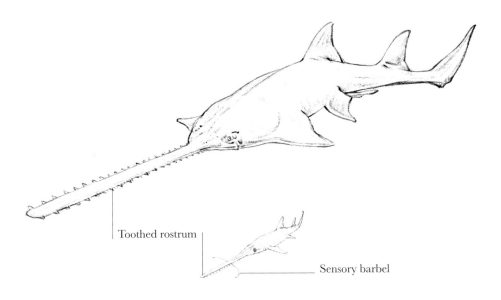

SAWFISH AND SAWSHARK
*Despite both wielding a toothed rostrum, sawfishes (top) are much larger
and lack the distinctive sensory barbels of the true sawsharks (bottom).*

Fight or flight

In the life and death struggle to escape predators, some oddball sharks have evolved very creative evasive maneuvers. For example, Swell Sharks (*Cephaloscyllium ventriosum*) are so called because they can puff themselves up to double their normal size. As these sharks are quite small—maxing out at 3.5 ft (1.1 m)—this is a pretty ingenious, one might even say "swell," adaptation. When it feels threatened, the Swell Shark will bend its body back on itself to grab its tail with its teeth and then suck large amounts of seawater into its stomach. Similar to a pufferfish, this makes the little shark look bigger and more intimidating. Additionally, if this happens while it is hiding in a rock crevice, an erstwhile predator would be advised not to waste energy attempting to unwedge the Swell Shark.

Other sharks take a different approach. Consider the Puffadder Shyshark (*Haploblepharus edwardsii*), which is also more colorfully known as the Happy

PUFFADDER SHYSHARK DEFENSIVE STRATEGY
*Puffadder Shysharks curl up and cover their face with their tail
when they feel threatened.*

Eddie Shyshark. These sharks earned their common names because they and their close relatives will curl their bodies up like a donut and hide their faces under their own tails when threatened. You might assume this to be an ineffective "if-I-can't-see-you-then-you-can't-see-me" strategy, but in fact, scientists suspect it makes these little sharks difficult for larger predators to swallow.

Slingshots

Undoubtedly one of the most bizarre-looking of all sharks is the very appropriately named Goblin Shark (*Mitsukurina owstoni*). These sharks are ambush predators, remaining very still near the ocean floor until potential prey, such as a small fish, squid, or crustacean passes close by. As we have discussed previously, the deep sea is a relatively food-poor space, but natural selection has ensured that a predator like our Goblin Shark can survive by not allowing food to escape at any cost. They could have evolved the physiological machinery and body shape of an aquatic sprinter to accelerate rapidly and catch that fish, but evolution opted for a different pathway, one that saves the energy required to move the shark's entire body. Instead, the sluggish Goblin Shark slingshots their entire jaw out of their mouths. To do so required evolving a completely unique facial morphology. As the ligaments in the jaws of many sharks are somewhat flexible, it is not uncommon for other species of sharks to protrude their jaws somewhat, but the Goblin Shark takes this trick to a whole new level. Their jaws are nearly ten times more protrusible than other species and can extend as much as nine percent of the total body length of the shark. Rather than being

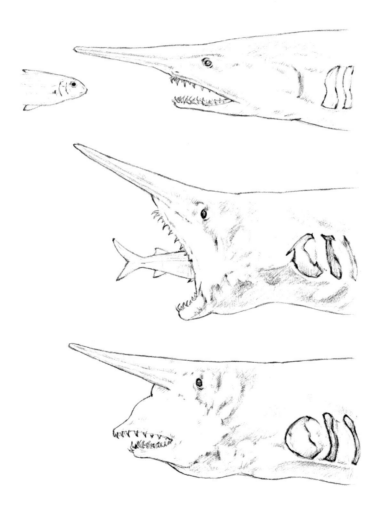

GOBLIN SHARK SLINGSHOT FEEDING
Thanks to their specialized elastic ligaments, Goblin Sharks can spring their jaw, as far as 9 percent of their total body length out of their mouths, to feed.

fused to their skull, their powerful jaws are suspended by two pairs of elastic ligaments. When the jaws are in their resting position, these ligaments are pulled very tight, but when the shark bites, the ligaments relax their tension and catapult the whole jaw forward like a rubber band, at speeds of 10 ft/3.1 m per second. Not surprisingly, this is described as "slingshot feeding."

These prehistoric-looking beasts were featured in a 2009 B movie, but contrary to the plot of *Malibu Shark Attack*, Goblin Sharks are neither extinct nor dangerous, and they are not found along shallow coastlines but rather in the deep sea to about 3,000 ft (900 m). They can exceed 10 ft (3 m) in length and, notwithstanding the movie and some other depictions, they do not swim with their jaws protruding.

Feeling flat

As we mentioned in Chapter 3, several groups of sharks seem to have missed the memo that they should have diverged from the rays and possess a more batoid-like, flattened body shape. Angel sharks (order Squatiniformes) and wobbegongs (family Orectolobidae) both have laterally expanded pectoral fins which have fused to their flanks near their heads, similar to skate wings, but they also have strong, muscular tails like other sharks. So, they almost look like an evolutionary stepping stone between the two lineages. But this odd body form is what makes them so perfectly suited to their lives as benthic ambush predators lying in wait for prey on the seafloor.

Also known commonly (but mistakenly) as monkfish (and sometimes so labeled in fish markets), there are at least 23 species of angel sharks, but new species are being discovered very regularly of late. Interestingly, their genus name, *Squatina*, comes from the Latin meaning "skate," so it seems the ancients had some trouble telling the flat sharks and batoids apart as well. Found worldwide in temperate and tropical waters, usually in the shallows, they are medium-sized sharks that generally peak at around 7 ft (2.1 m). At least 12 species are currently listed on the as Endangered or Critically Endangered.

Wobbegongs are predominately found in the western Pacific and eastern Indian Oceans, off Indonesia and Australia. But one species, the Japanese Wobbe-

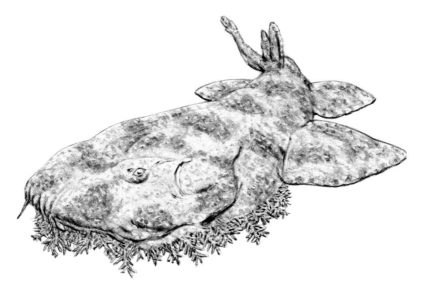

TASSELED WOBBEGONG
*Its patterned skin and tasseled snout are perfect camouflage
for a wobbegong lying in wait for its next meal.*

gong (*Orectolobus japonicus*), lives farther north, off the coast of (you guessed it) Japan. They are mostly less than 4 ft (1.25 m) long, but the Banded Wobbegong (*O. halei*) can exceed 6.6 ft (2 m). All 12 species of wobbegongs sport fantastic and frankly odd dermal flaps around their mouths that help them to camouflage against the seafloor. This feature is what inspired their common name, which in the Aboriginal Australian means "shaggy beard." These projections are most pronounced in the aptly named Tasseled Wobbegong (*Eucrossorhinus dasypogon*), which is especially frilly. Like many other carpet sharks, all wobbegongs have beautifully patterned skin, with markings and colors unique to each species. When they are lurking amid corals, these patterns make them very difficult to spot, so they can lunge out at unsuspecting fishes, octopuses, crabs, and lobsters.

Angel sharks also camouflage well, but they favor flat, sandy areas between rocks, where they can target small fishes, crustaceans, clams, squids, and gastropods. These sharks are especially unusual because they have an upside-down, or hypocercal, tail. In most sharks, the upper lobe of the asymmetrical caudal fin is larger than the lower lobe. This is known as a heterocercal tail. In the five

163

ANGEL SHARK GENERATING LIFT FOR AN AMBUSH
*Angel sharks have a unique upside-down tail, with a larger lower lobe compared
to the upper, to help generate lift as they spring off the sea bed to ambush prey.*

species of sharks in the family Lamnidae, which includes the White Shark and Shortfin Mako (*Isurus oxyrinchus*), the top and bottom lobes are closer to similar in size: a homocercal tail. However, only in the angel sharks is the lower lobe of the tail the larger of the two. We don't see the hypocercal tail in any other groups and it has developed because it allows an angel shark to gain lift when propelling itself from the seafloor to ambush prey. The more conventional shark tail shape pushes the head downward and so species with a heterocercal tail have developed larger pectoral fins to generate lift.

Watch me whip

Speaking of shark tails, we would be doing readers a disservice if we omitted the three dazzling species of thresher sharks: the Common Thresher (*Alopias vulpinus*), Bigeye Thresher (*A. superciliosus*), and Pelagic Thresher (*A. pelagicus*). These sharks all have an elegant, elongated upper caudal fin which can be as long as the rest of the shark's body. In addition to its role in propulsion, this oddball feature plays a vital role in hunting, as thresher sharks will use their tail to herd

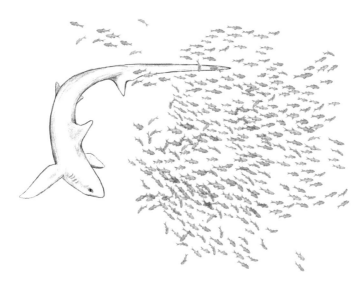

THRESHER SHARK "TAIL SLAP"
The three species of threshers have evolved an extra elongated upper caudal lobe of their tail, which they can whip through the water to stun their prey.

and tighten schools of prey (like anchovies, herring, and squid) before swatting, stunning, and eating them. These "tail slaps" may occur over only 33 milliseconds—that is just 0.033 seconds. Reaching breakneck speeds of nearly 50 mph (80 km/h), they can cause such extreme decreases in water pressure that air bubbles may form in the water. Threshers also swat the surface water, or perhaps swat fish there, which likely explains reports by commercial fishers who have caught threshers hooked by the tail, presumably as they swatted the bait. You need not fear threshers. As well as being one of the small number of oceanic species that most of us will never encounter, the jaws of thresher sharks are very small and weakly calcified, reflecting their small, soft-bodied prey.

Packing heat

Most people correctly presume that fish are cold-blooded as a group, or more correctly, ectothermic. Ectotherms are animals whose body temperature is determined primarily by that of the environment. All invertebrates, and most bony fishes and chondrichthyans, are ectotherms. The opposite condition, endothermy—the ability to elevate the body temperature above that of the surrounding environment—is found principally in birds and mammals.

It is especially difficult for an aquatic organism to trap heat in its body because the physical properties of water make it a thermal thief and thus the heat that every animal produces as a by-product of metabolism is mostly absorbed by the water in which they live. If you have gills, heat is expeditiously lost there, since their architecture requires the metabolically warmed blood to be separated from the watery environment by only the most minute of distances, so the gills can fulfill one of their core functions of absorbing oxygen. The presence of lungs to breathe air as opposed to gills, along with insulation provided by blubber, fur, or feathers, allows aquatic mammals and birds to be endothermic.

However, nature frequently disobeys the rules, and so some 450 MYA several fish lineages independently evolved the machinery to be endothermic. It is rare, though, and found only in four families of bony fishes and a few species of cartilaginous fishes. The Lamnidae (five species of mackerel sharks), Alopiidae (all three threshers), and Mobulidae (at least two species of mantas) have evolved ways of trapping heat to elevate their body (or specific parts of their body) temperatures by as much as 18°F (10°C) higher than that of the surrounding water. Very recently, scientists have also discovered that Basking Sharks (*Cetorhinus maximus*) and Smalltooth Sand Tigers (*Odontaspis ferox*) also possess the morphology required for this feat.

Unlike other endotherms (including us) that warm their entire body, all endothermic fishes are more correctly referred to as mesotherms (meaning "intermediate heat" in Greek), since they warm only certain parts of their bodies. This phenomenon has also been called regional endothermy. The threshers heat their brain and eyes, Basking Sharks warm their muscles, and the more advanced species—for example, the mako sharks—also warm their stomachs and guts.

Mesothermic sharks have an exceptional ability to retain metabolic heat which has already been produced as a by-product of locomotion and digestion. They effectively recycle all their metabolic heat back into their bodies. This is possible thanks to a system of countercurrent heat exchangers in the circulatory system, called the retia mirabilia (meaning "wonderful nets"). Basically, blood vessels whose blood has been warmed at the body core are juxtaposed with vessels containing cooler blood from the periphery of the body, and the heat diffuses from the warm blood to the cooler blood before it can be lost to the environment.

Controlling their body temperature at least somewhat independently of the surrounding water temperatures means that all mesothermic sharks can tolerate colder waters better than ectothermic species. This increased tolerance may have contributed to their expanding into higher latitudes as well as into colder depths without having to compromise how active they are, a phenomenon known as thermal niche expansion. As we mentioned earlier, their wide thermal tolerance may be what allowed the White Shark to out-survive Megalodon (*Otodus megalodon*).

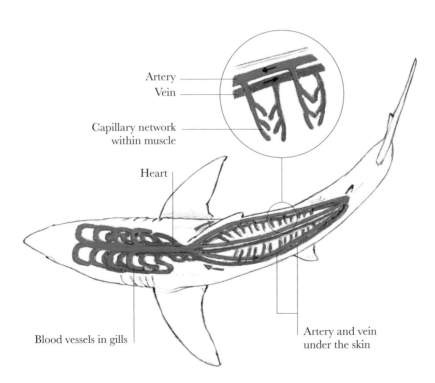

Artery

Vein

Capillary network
within muscle

Heart

Blood vessels in gills

Artery and vein
under the skin

COUNTERCURRENT HEAT EXCHANGE SYSTEM
*Some sharks in the order Lamniformes have evolved specialized
countercurrent heat exchangers that cycle heat back into the body, to elevate
the temperature of the brain, eyes, muscles, and gut.*

In the species with more advanced mesothermy, the ability to maintain their high metabolic rate even in cold water means these sharks have extremely efficient digestion and explosive muscle contractions. This is because a body that is warmer than its environment can enable the chemical reactions underlying these physiological processes to be much faster. The upshot is that they have *a lot* more energy for swimming, so this group includes some of the most active and dynamic of all shark species. The Shortfin Mako is the fastest shark in the sea—capable of bursts of 45 mph (5 km/h). They can also jump higher than any other shark, breaching up to 30 ft (9 m) out of the water. Even White Sharks as large as 21 ft (6.5 m) and weighing perhaps 4,400 lb (2,000 kg) are capable of launching their entire bodies out of the water when hunting. The downside is that this metabolic machinery must be constantly fueled, so these sharks require a lot of food.

A VARIED DIET

It has always been undisputed that sharks are carnivores—with gut biochemistry and physiology adapted to digesting only animal prey—and do not require other food groups to get all the vitamins and minerals their body needs. However, scientists have recently discovered that this may not necessarily always be the case. Researchers studying Bonnethead Sharks (*Sphyrna tiburo*) off the coast of the United States have learned that, while these sharks primarily eat mollusks, crustaceans, and cephalopods, they may also be able to digest seagrasses.

At first the experts wondered if the Bonnethead Sharks might be consuming plant material accidentally while foraging or were eating it to protect their stomach lining from the spiny shells of their favorite prey, the blue crab. However, these sharks are apparently capable of digesting fibrous plant material, thanks to specialized carbohydrate-degrading enzymes in their gut. Additionally, analyses of their tissues have revealed that Bonnethead Sharks assimilate plant materials into their bodies as they grow, and seagrass is, in fact, an import-ant component of their diet at all different life stages: from newborn to mature adult. This means the Bonnethead Shark is the first and one of only two sharks thought to be omnivores. The idea of a shark being capable of digesting plant material so challenges the paradigm of their carnivorous nature that numerous shark scientists seriously question this discovery.

Filter function

Whether they take a high octane or lazy approach, as a group sharks are top predators. However, there are a few species that feed much lower in the food chain, so much so that you might not have even realized they were predators. But prey size does not matter when it comes to being considered a predator. Indeed, the Whale Shark (*Rhincodon typus*), Basking Shark, and Megamouth (*Megachasma pelagios*) are all predators, despite being filter-feeders. This means they are able to reach their incredible sizes of 61 ft (18 m), 40 ft (12 m), and 17 ft (5 m), respectively, fueled mostly by small, even microscopic planktonic organisms. Rather than ambushing or giving chase to large prey, these sharks filter diatoms, protozoans, and small crustaceans, squid, and fish, as well as the eggs and larval stages of larger animals, out of the water column to eat. They must consume hundreds of pounds of these little morsels—numbering in the millions of individuals—every single day. This is possible thanks to specially adapted gills that act as a feeding sieve. In these species, specialized cartilagi-nous structures called gill rakers project out from the gill arches to form a fin-gerlike mesh that traps small animals inside the shark's mouth as the water flows back out the gill slits. As a result, these sharks have no use for their teeth at all and these are actually vestigial (a functionless body part). Whale Sharks are part of the order Orectolobiformes, whereas Megamouths and Basking Sharks (and the ancient species of soaring shark called *Aquilamna* that we men-tioned in Chapter 2) are all Lamniformes. This means that filter-feeding has evolved several different times across the shark lineages.

Basking Sharks may be the most iconic filter-feeding shark, as they are so easily recognizable thanks to their cavernous gape. Their mouths can be as

BASKING SHARK FILTER-FEEDING
*Basking Sharks are one of only three species of
sharks that are filter-feeders.*

wide as 3 ft (90 cm) across when open. They feed by moving slowly forward near the surface, so that water (and all those tasty microorganisms) flow into their mouths and back out their gills. This is known as ram filtration. They can filter through over 2,200 US tons of water per hour this way. That's enough to fill an Olympic-sized swimming pool.

Whale sharks filter-feed in different ways. They can choose to use ram filtration or feed by a process known as suction feeding. Through this method the Whale Shark can stay completely still and draw water and food in by opening and closing their mouth to create suction.

Megamouth Sharks take stationary feeding even further. A distant cry from the sleek, muscular, predatory sharks, Megamouths have a somewhat flabby body. They move so slowly that they have developed a system to lure food into their mouths without ever having to move at all. This is possible thanks to a bright white stripe of tissue around their top lips. Reflecting what tiny amount of light that can penetrate through the gloom of the deep, this lip stands out starkly and attracts in prey. Despite being huge, as they live at incredible depths, Megamouths have rarely been observed or caught. In fact, their sightings only number in the hundreds since records began, so they are still incredibly mysterious.

Twinkle, twinkle, little shark

While it has been disproven that Megamouth's lips produce light to lure prey, some species of sharks can, in fact, glow in the dark. Being able to make light (known as bioluminescence) or transform light (biofluorescence) is quite common for animals living in the deep ocean, but it is not widespread among sharks. As many as 20 deep-sea sharks are capable of glowing. The ability is not restricted to one family of sharks and has evolved multiple times along the shark evolutionary line. Some species of wobbegongs (family Orectolobidae) and a few catsharks (family Scyliorhinidae) can glow in the dark, but the ability is most common among the dogfish (order Squaliformes), where three families—Dalatiidae (kitefin sharks), Somniosidae (sleeper sharks), and Etmopteridae (the aptly named lantern sharks)—all put on a light show.

Some marine animals glow thanks to symbiosis with another species. This means a larger animal, like a squid, plays host to another organism, like a bacterium, that is able to create light. Sharks are different. Some species living at shallower depths are able to transform and reproject photons from sunlight to glow, but deeper sharks produce light thanks to fluorescing proteins that they make in specialized organs within their skin called photophores. These cup-shaped structures contain pigmented cells which are capable of biochemical reactions that produce blue-green light (with a wavelength of 455 and 486 nanometers). In some species, specialized lens cells or reflector molecules sit on top of the whole structure, to focus and enhance this light. Sharks are also among the only animals that can control their glow via hormones. The production of melatonin (among others) allows light production to be switched on and off as needed.

Bioluminescence in sharks is unique to each species because its function varies depending on how that species lives, how it behaves, and what it eats. For example, the Chain Catshark (*Scyliorhinus retifer*), a relative of our Smallspotted Catshark (*S. canicula*), glows with intricate networks of patterns that match the markings of pigmentation in the skin. This is a form of visual communication because it allows the shark to recognize other individuals of their species in the deep, dark ocean, where it might otherwise be difficult to tell them apart from other similar-looking sharks. This can be especially useful when looking for a mate. Swell Sharks have gone one step further and their patterns of light are sexually dimorphic, so that individuals can discern males from females amid the gloom.

Some sharks use their light to aid in hunting their prey. The largest of all glowing sharks—in fact, the largest of all glowing vertebrates—is the Seal Shark (*Dalatias licha*), which reaches some 4.6 ft (1.4 m) in length. These sharks glow only on their ventral side and are thought to use this down-light to illuminate the ocean floor as they search for food. On the other hand, some sharks use their bioluminescence for camouflage—to avoid predators. This sounds counterintuitive, doesn't it? How could emitting bright light in a dark environment help you hide? Wouldn't that be a neon sign advertising your presence? In fact, it depends

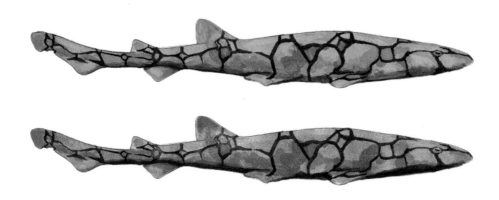

BIOLUMINESCENT SHARKS
To help them find prey or a mate, several families of sharks have evolved unique patterns of biofluorescence. The Chain Catshark is shown.

172

on your perspective. Many predators in the deep, dark oceans search for prey by looking upward. This is because in very low light it is easiest to spot the dark silhouette of prey contrasting against the light coming down from above. As they only glow on their bellies, Seal Sharks actually become invisible in the water when viewed from below. This is known as counterillumination.

Probably the most impressive bioluminescence can be found in the American Pocket Shark (*Mollisquama mississippiensis*). As well as having fluorescing skin, this shark also squirts pockets of glowing liquid into the water as a defensive strategy to confuse predators—like a flashy, glowing squid.

Smart cookie

The Cookiecutter Shark also glows, using its light-emitting belly and black collar to break up its silhouette, so predators can't spot it from below. But, amazingly, this is not the most incredible feature of these bizarre little sharks. In fact, they might supplant the White Shark as the most fearsome shark in the ocean if they were bigger and found in shallow water. As its sausage-like body only grows to about 1.8 ft (50 cm), you'd think the Cookiecutter Shark would know its place in the world and eat small fish and squid. But this beast punches above its weight, targeting much larger fish (including sharks) and marine mammals like whales and dolphins. There are even documented cases of them preying on humans and—you're not going to believe this—nuclear submarines and other military equipment. In fact, it was the US Navy that discovered and described the species, after they had continuously caused such distinctive damage to their equipment.

So how does this small animal take on such big prey? Well, Cookiecutter Sharks are a type of ectoparasite, taking bites out of larger animals and leaving them alive. This is thanks to a dark collar of pigmentation around their throats—the only part of their skin that does not glow in the dark. When viewed from below by a predator in search of prey, these markings resemble a little fish. Tricked into thinking the shark is a smaller animal—a free lunch, so to speak—a swordfish, tuna, or dolphin will approach. When the predator investigates, the Cookiecutter Shark, again invoking the notion of the scarcity of cost-free midday meals, turns the tables and accelerates toward it, using its relatively large and powerful caudal fin, and latches on with its sharp, triangular teeth. Their specially adapted fleshy lips create suction on the larger animal's flesh and they then spin while holding on with their powerful jaws, carving out a chunk of meat or blubber. This leaves very distinctive cookie-sized craters behind, which gave these sharks their name.

True colors

Some sharks are actually able to subtly change the color of their skin. For example, scientists studying White Sharks often witness an animal switching color throughout their observations over the course of a day or several weeks. This is possible thanks to hormones, such as adrenaline, stimulating specialized cells in the skin called melanocytes. When these cells contract and relax, they change color from dark to a lighter shade and back again. This means the skin on a White Shark's back can change through the grayscale spectrum, providing the shark with better camouflage. This might not be as dramatic as the rainbow shows put on by chameleons and octopuses, but it's pretty handy for a White Shark to camouflage itself when sneaking up on seals.

Scientists have also now discovered that some sharks actually get a suntan. Young Scalloped Hammerheads (*Sphyrna lewini*) become progressively darker on their backs when exposed to higher levels of ultraviolet light in shallow waters, compared to when they live at greater depths. Scientists suspect these tans protect the shark's skin from sun damage to some degree. This is very rare among fishes and has been documented in only a handful of species.

Other sharks will drastically change colors or pigmentation patterns as they age. The stripes of the Tiger Shark, for instance, are more pronounced when they are young, but fade with maturity. However, there is no species that takes this to such an extreme as the Zebra Shark (*Stegostoma fasciatum*). At birth, they sport the distinctive vertical black stripes that earned them their name. However, as they age, the Zebra Shark's dark saddles break up into black and brown patches, revealing more yellow pigmentation, so by the time they are fully mature, they are spotty like a leopard. The changes are so pronounced that the scientists that first described these sharks classified young sharks, juveniles, and different-colored adults as distinct species before modern DNA analyses finally revealed the truth.

These changes are no accident. Counterintuitively, the bold stripes protect young Zebra Sharks from predators in several ways. Firstly (for the same reason that drives this adaptation in terrestrial zebras), the patterns allow them to blend into a crowd of other Zebra Sharks and confuse predators. This is known as the predator dilution effect. Secondly, as the patterns mimic those worn by venomous sea snakes, they look like a much more dangerous animal and this can repel predators from even wanting to try to eat the little shark in the first place.

Buoyed up

As we seen in Chapters 1 and 3, the lack of a swim bladder in sharks has drastically altered how they live. Instead of floating in the water like most bony fish, the majority of pelagic sharks must swim in order to keep their heavier-than-seawater bodies from sinking. Benthic sharks have opted to go with gravity and loll around on the seafloor. However, a few oddball species have come up with some novel ways to counter their negative buoyancy.

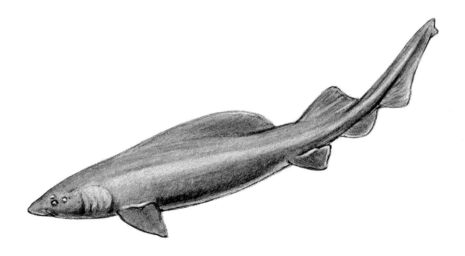

FALSE CATSHARK
Deep-sea False Catsharks have an elongated first dorsal fin, almost like an eel.

Take as an example the False Catshark (*Pseudotriakis microdon*). Recognizable thanks to a bizarrely elongated, keel-like first dorsal fin, which is similar to that of an eel, this deep-sea species lives as deep as 4,500 ft (1,400 m). But the False Catsharks did not sink there. Unlike most other sharks, they are almost completely neutrally buoyant. Thanks to an enormous oily liver, which can make up as much as a quarter of their body weight, they are able to hover in the water column. This is perfect for a sluggish, somewhat flabby creature that only very rarely puts on short bursts of speed to snatch at prey.

On the other hand, the sleek, intimidating-looking Sand Tigers (*Carcharias taurus*) have gone down a different route: they will actively expel gas to control their buoyancy. If they want to remain shallow, Sand Tigers will increase their buoyancy by gulping air from the surface and holding it in their stomach—like inflating a Buoyancy Compensator Device (BCD) when SCUBA diving. Just like a BCD, if you want to go deeper, you have to release some air … When Sand Tigers must get rid of excess air in order to descend, they do so by farting. Charming.

9

A TALE
AS OLD
AS TIME

GODS, GUARDIANS, AND GUIDING SPIRITS

When you consider that sharks have been roaming the oceans for hundreds of millions of years, it becomes apparent that human beings constitute only a minuscule, even insignificant, chapter in the shark story. Sharks are ancient and we are the new kids on the block. Yet, from our perspective, our relationship with sharks is a tale as old as mankind itself. Whether the earliest humans encountered sharks when fishing for food or after taking to the high seas to migrate to new lands, archeological evidence shows us that our lives have been intertwined with sharks for as many as 300,000 years.

Our ancient ancestors told stories about sharks to help them understand their role in a confusing and often frightening world. Thus, there is a rich history of myths, legends, folklore, and theology involving sharks from all over the world. The perception of sharks held by these ancient peoples was complex and conflicting. Some worshipped and celebrated sharks, others feared and demonized them. Sharks have variously been considered tricksters, villains, and guardians. They have been both worshipped as gods and feared as demons, venerated as protectors, and blamed for causing catastrophes. The shark might be the form taken by a benevolent spirit on the one hand, or the monstrous shape assumed by a person transmogrified as punishment for committing terrible crimes.

Given the prevailing myth of contemporary Western culture that paints sharks as villains and mindless eating machines, it may be surprising to discover that throughout human history, it is far more common for sharks to be venerated and respected than to be feared and hated. Historically, sharks have been respected as powerful forces of nature; esteemed as the ghosts of ancestors or lost children and as the carriers of transmuted human souls; perceived as protectors, guardians, helpful spirits, and avengers of justice; and worshipped as gods and agents of deities. In these tales, the sharks are the good guys, and they have been honored with offerings, temples, and totems. Although we find these stories fascinating, we'll consider only a very small sampling, with apologies to the rich heritage we must overlook.

Forces of nature

Some of the richest and most complex mythologies featuring sharks come from the Indigenous peoples of Hawaii. For centuries, sharks were considered more than just animals in Hawaiian culture; they were respected as powerful forces in marine ecosystems and revered as protective spirits and even gods, nine of which feature in Indigenous Hawaiian theology. Polynesian cultures respected the majesty of sharks in the water—their habitat, not ours—and these peoples did not attribute any malice to the random attacks on humans (a conclusion contemporary humans might also have reached in the absence of omnipresent sensationalized accounts in the media). Sharks were simply regarded as powerful animals deserving of respect. In Fiji, for example, pearl divers would attend ceremonies with shark charmers before daring to head into the water.

PEARL DIVER
*Pearl hunters free dive in shark habitat, a tradition that dates back
thousands of years.*

Spirit hosts

In several cultures, sharks were revered as the hosts of spirits. For instance, in Papua New Guinea, they were considered the embodiment of the ancestors. On their deathbeds, Papuan people would proclaim their intentions to become a shark after shedding their human form, and if a shark was noted for its exceptional size or color, or if it seemed to stick around a meaningful location, residents believed the shark to be the embodiment of a ghost. Offerings of food would be made to the shark and the animal would be called by the deceased's name.

Assigning human names to other animals may be interpreted as anthropomorphism, but there is a lesson here: *We value what we name.* Shark researchers who tag their animals, as well as shark ecotourism operations, frequently name the sharks they encounter, and people form emotional attachments to these named sharks. We called a Blacktip Shark (*Carcharhinus limbatus*) that raced between fishing piers on the coast of South Carolina, in the United States, Ricky Bobby, and a Sandbar Shark (*Carcharhinus plumbeus*), Rachel Carson.

Sharks have also commonly been venerated with totems—objects of spiritual importance—with wooden carvings of sharks displayed at temples and shrines around the world. In the Marshall Islands, in the central Pacific Ocean, no animals were more significant symbols than sharks, and each tribe celebrated its own special shark. If someone from another group insulted their shark or killed one, the tribe would even be willing to go to war over it.

There are also many stories across myriad cultures where sharks have been seen as helpers. For instance, in the Solomon Islands, in the South Pacific, it was believed that good sharks helped to protect people when they were fishing or swimming. In Tahitian theology, the god Tawahaki was transported across the sea thanks to a powerful charm that allowed him to ride on the back of a helpful shark. There are also several more recent myths where sharks have acted as helpful agents. A Mediterranean legend from 1573 tells of a ship in trouble at sea being saved from sinking by a sawfish that wedged itself into the hole in the hull, corking the leak. The crew believed their devotion to God brought this helpful ray to the rescue and the rostrum of this heroic sawfish is now displayed as an important Christian relic in the Basilica del Carmine Maggiore, in Naples, Italy. A twentieth-century myth from Australia recounts how controversial mining magnate Cecil Rhodes was able to make his fortune by receiving stock market updates before everyone else because the *Financial Times* was rapidly transported to him from London to Australia after being eaten by a shark. There are also many different examples of sharks as guardians in different mythologies. Throughout Polynesia, islanders would seek the protection of their shark gods before undertaking voyages at sea. In traditional Fijian culture, legend tells of a shark that defended the reef entrance to the islands.

Sharks are still revered as important spiritual animals in many contemporary cultures. In the simplest form, sharks are honored with images on different countries' currencies. Throughout the Pacific, many communities still find their traditional stories to be very relevant and continue to recognize sharks as helpful spirits. Modern surfers and divers (perhaps including you) who wear a shark-tooth necklace in the water unwittingly also honor shark mythology, because this tradition dates back to ancient Hawaiian culture. In these legends, lord of the seas Ohav-Lai went to battle in the depths of the Pacific to earn his supremacy. During the day-long fight he pulled out a shark's tooth and, upon emerging from the fight victorious, he strung the tooth on a necklace. To this day, some divers in the South Pacific will enter the water only if they are wearing their shark-tooth necklace, which they believe will protect them from sharks.

Gods and deities

Sharks also often feature as agents of the divine in ancient theologies. For example, Mortlock Islanders would call on their good spirit named Ulap, who had dominion over many marine animals, including sharks. In one story, Ulap answered the prayers of those who had been capsized from their canoe, bringing forth an enormous shark that scared off any dangers and ensured they made it safely back to shore.

Sharks have also been worshipped as gods themselves in many cultures. Gilbert Islanders celebrated a deity by the name of Tabaruaki and in Tonga a deity named Taufa was revered, both of whom would often assume the form of a shark. The fearless Fijian god Dakuwaqa (also known as Avatea in the Cook Islands and Takuaka in Tonga), who is depicted as a man with human legs, but the head of a shark, far from being a monster, helps fishermen avoid dangers at sea and protects people from other sea demons. Many people revere Dakuwaqa to this day.

MIGHTY FIJIAN SHARK GOD DAKUWAQA
Also known as Avatea in the Cook Islands and Takuaka in Tonga,
Dakuwaqa is depicted as a man with human legs, but the head of a shark.

THE DEVOURING MONSTERS

There are also numerous stories where sharks have been portrayed as demons, monsters, or fearful gods. While less common than tales of sharks as benevolent beings, these beliefs and stories are as diverse and fascinating as the cultures themselves. The Mayans, for example, quivered before the great demon shark Chac-uayab-xoc. In Japan, a shark-man was once believed to whip bad weather into typhoons. The west African tribe known as the Bavili referred to sharks as *Nquimike Ku Vuka*—"those that devour." In Tahiti, the deities Tauraati and Ruahatu employed sharks as their agents of divine vengeance. An Aboriginal group of Australia told the story of a great, bloody fight between a mythical dolphin being and a giant Tiger Shark called Bangudya, as a way to explain why the rocks they saw at nearby Chasm Island were so bloodred. One of the most famous fables about fearful sharks, from Brazil and Guyana, describes how a character called Nohi-Abassis attempted to encourage a shark to eat his hateful mother-in-law. When he was caught in this misdeed, the shark turned on him and amputated Nohi-Abassis's leg. The lesson: be nice to your mother-in-law.

In some cultures, people were even sacrificed to sharks in rituals or religious ceremonies. For instance, the people of the Solomon Islands once believed that angry sharks could be appeased through human sacrifice, so they would cast people—both living and dead—into the seas to feed them. In Hawaii, kings and priests would throw a noose among their attendants and the poor soul unlucky enough to be caught by it would be cut into pieces to feed and abate the sharks.

Unflattering representations of sharks are also prevalent in Greek mythology. In one story, the sun god Apollo punished several politicians who had offended him by sending a shark to attack them as they bathed. The story of Lamia has even changed the way scientists talk about sharks today. In this myth, the daughter of the ocean god Poseidon had an affair with Zeus, the king of the gods. His spiteful wife Hera kidnapped and murdered Lamia's children

SALMON SHARK
The genus name Lamna *for Salmon Sharks comes from a Greek myth
about a vengeful, devouring sea monster.*

in vengeance, driving the bereaved mother insane. Lamia then morphed into a giant, ravenous shark monster and endlessly devoured the children of other mothers in revenge. Lamia's name inspired the genus name *Lamna* for Salmon Sharks (*L. nasus*) and Porbeagles (*L. ditropis*) and also the order name Lamniformes. This scientific name literally means "insatiable devouring monster." How flattering …

FEEDING FRENZY?

The Western world's perception of sharks as monsters began with their biased representations in Greek mythology and our fears have subsequently been enflamed by continuous misrepresentations throughout our culture. As far back as 492 BCE, Herodotus wrote of hordes of sea demons devouring shipwreck victims in his *Histories*. Violent, terrifying shark attacks were often featured in important works of art, such as John Singleton Copley's *Watson and the Shark* (1778) and Winslow Homer's *The Gulf Stream* (1899). Sharks were the monsters of classic literature too. In *Twenty Thousand Leagues Under the Sea* (1870) Jules Verne described sharks as "iron-jawed" leviathans and Herman Melville devoted a full chapter of *Moby Dick* (1851) to gleefully describing a righteous "Shark Massacre." Our feelings toward sharks are thus no recent development.

Yet of late, negative perception of sharks has been ever more influenced by their presentation in the media. Sharks are generally in the news only when they are implicated in attacking a human. In movies or television shows they are often presented as archetypal horror movie monsters. Sharks are stylized as villains, rogues, a scourge—mysterious, man-eating denizens of the deep. This disdain and fear are now so deeply ingrained that the word "shark" has become synonymous with evil. There is no question in your mind what kind of a character a "loan shark" or a lawyer described as a "shark" will be, is there?

Despite the rarity of fatal bites, sharks have been depicted in the news as a huge threat. Even before TV and the internet, people were made aware of the horrors of shark attacks, and these reports were often sensationalized and disproportionate, even more so when you consider the other global tragedies occurring at the time. For instance, a spate of fatal attacks attributed to one shark dubbed the Matawan Shark in New Jersey waters in 1916 were widely and sensationally reported in the news for many weeks. Comparatively little coverage was given to the deadly polio epidemic sweeping New York and killing an average of one child per hour at the same time! While it is terrible for anyone to lose their life to a shark, when we put these losses into perspective, we start to wonder why we are focusing disproportionately on sharks. Did the Matawan Shark really pose such an enormous threat when, in the same week, over 19,000 British soldiers were lost at the Battle of the Somme?

There is also a history of scapegoating sharks amid maritime disasters. The news avidly reported on the frightful assaults sharks perpetuated upon sailors left in the water after the sinkings of the UK's RMS *Laconia* and RMS *Nova Scotia* in 1942, and the USS *Indianapolis* in 1945 (which is famously described in the movie *Jaws*). News surrounding the USS *Indianapolis* case paved the way for systemic scientific studies of sharks, funded by the US Navy. However, the truth is that the vast majority of the victims of these dreadful disasters likely died as a result of hypothermia, drowning, or infection, though there's no way of knowing for sure.

Investigations into the patterns in contemporary news coverage of sharks have clearly demonstrated a significant anti-shark bias. A scientific study assessing global reporting on sharks between 2008 and 2017 found that, among the 1,800-plus features, a reference to the *Jaws* franchise was mentioned in 7.5 percent of articles about sharks and a fear of sharks was generally included in 15.9 percent of reports, even those focusing on shark conservation. Similarly, a study of Australian and US media coverage determined that, of 300 news articles between 2000 and 2010, more than half discussed shark attacks, while only 11 percent talked about shark declines. The study also uncovered many inaccuracies and outright errors in the articles.

Not only has contemporary journalism focused uneven attention on the risk that sharks pose to human beings (as opposed to the threat we are to them), they have also contributed to widespread public misunderstanding about sharks, and this has not been harmless. Portraying sharks in a negative light can seriously limit public support for their conservation and can even affect policy decisions of world governments.

The *Jaws* effect

Let's be honest, the first thing that comes to most people's minds when they hear the word "shark" is the movie *Jaws*. This is so iconic that it is almost impossible not to hear people singing the ominous two-toned theme tune when they see sharks in aquariums or on diving trips, no matter where in the world you happen to be. When Spielberg's movie adaptation of Peter Benchley's novel hit the big screen in 1975, the vast majority of the general public scarcely had any interactions with sharks or knew much about them at all. But on the release of *Jaws*, sharks exploded into the mainstream media and almost overnight sharks were thrust into the public consciousness, instantaneously transformed from a somewhat distant, curious ocean-dweller into a malicious movie villain.

We would be remiss if we failed to acknowledge the undisputed shark media juggernauts, Discovery Channel's *Shark Week* and National Geographic's *Sharkfest*. These saturate cable airways in over 70 countries for a week or more each summer and in countless replays, with recent annual viewership exceeding 50 million. These shows have elevated shark appreciation. At the same time, however, many of the episodes do just the opposite, in their attempt to entertain more so than educate. The worst of these validate irrational fears, provide dubious facts, may be totally fictional, misuse science, set up ridiculous scenarios, and/or waste resources, including viewers' time.

Whether you enjoy the movie or not, it is absolutely undeniable that *Jaws* was very bad for sharks (although as the movie has aged, the effect has been more mixed). After the release of the movie, tournaments and shark-fishing trips began to spring up all over the United States, with many different species of sharks, some not even remotely dangerous to human beings, being indiscriminately slaughtered. In the US, for example, over 25 million lb (11.4 million kg) of sharks were landed by recreational fishers in 1979 alone. This impact was so significant that scientists coined the term "The Jaws Effect" to explain how the blockbuster movies affected real-life sharks.

Since *Jaws* there has been an endless slew of unrealistic, overtly fantastical, and sometimes downright silly movies featuring sharks as the villains. Whether you find entertainment value in these movies or not, they represent sharks as ready to launch an attack on a human being at any time (even from the air) and thus perpetuate the stereotype that sharks are mindless killing machines. What's more, they also often imply that destroying sharks is the only path to protecting people, an attitude that deafens people to alternative, more truthful narratives about the plight of sharks and their ecological importance.

The sad truth is that this is an endless, self-perpetuating cycle. Sensationalized shark events in the news, in turn, make sharks more desirable for horror movies, which makes people more scared of sharks, which encourages the media to continue to show sharks … and the cycle goes round and around. Basically, scary sharks sell newspapers and tickets at the box office. Galeophobia (a fear of sharks) has become trendy; the media are selling it and we are buying it.

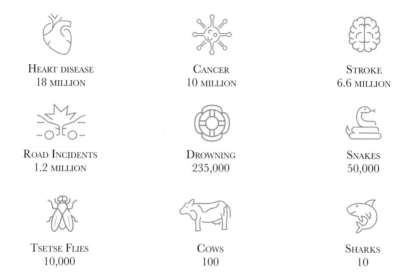

AVERAGE GLOBAL ANNUAL CAUSES OF DEATH
*Far more human deaths each year are caused by illness and accidents
than by shark attacks.*

MYTH BUSTING

In reality, sharks pose a minuscule threat to human beings. Only between six and ten people are reported to die as a result of unprovoked shark attacks annually, and in about 85 percent of shark bite incidents, the victim thankfully survives to tell the tale. This is not to downplay how tragic and traumatic these episodes are, but statistically you are more likely to be injured by your toilet or by an air freshener. You are far, far more likely to die as the result of an accident in your home and you are tens of thousands of times more likely to be killed in a car crash. Inhaling the dust emitted by tires and running shoes as they wear out is also a bigger risk to our health than sharks are. In the United States, air pollution causes as many as 200,000 premature deaths annually. Your local grocery store and its role in promoting unhealthy eating is also much more of a threat to your health (not to mention the very real chance of being run over by an SUV in the parking lot).

Sharks also pose much less of a risk compared to other animals; many more people die in attacks by wild animals like hippos and crocodiles, and many more are killed by domesticated animals like cows and dogs each year. We are also far more dangerous to each other than any shark is to us. You are statistically much more likely to be killed by another human being than by a shark. In fact (believe it or not!), for every one person bitten by a shark, there will be five incidents of people being bitten by New Yorkers.

In short, many people's perception of the risks that sharks pose is massively off. When asked if they are afraid of entering the water at the beach, many respond with an emphatic "yes," but if you were to ask us personally, we fear being decapitated by a jet ski, sliced by the fin of a surfboard, carried out to sea by a rip current, stung by a venomous jellyfish, or sickened by polluted water … but sharks as a cause of our fear? Not so much. There is also a pervasive (if wrongheaded) public perception that all sharks represent threats to people. You won't be harmed by a Smallspotted Catshark (*Scyliorhinus canicula*), for instance, or by any of the vast majority of shark species, unless you get nipped while removing a hook from one.

Despite the rarity of shark bites and attacks, many people also believe without verification that they are on the rise. But this conclusion is far from clear. The International Shark Attack File (ISAF)—an authoritative global leader in monitoring shark bites—reports that there has been an increase in bites in recent years, with 80–100 shark incidents (most not fatal) now being recorded annually. However, when we consider the absolute number of shark bites against the massive increase in how many people are entering the oceans, it would seem incidents may be decreasing. Bite rates in the top ten countries that regularly report the highest number of shark incidents (including the United States, Australia, and South Africa) have actually seen declines since 2016. Year-to-year variation in shark bites, or even longer trends, could be explained by a number of factors, none of which have been rigorously tested. These include better reporting of interactions; habitat alteration; increases in seawater temperatures associated with climate change, which may alter shark migration patterns; more people entering the waters; and the impacts on sharks' prey caused by overfishing. It is likely that the massive declines in shark numbers around the world have played a significant role as well.

In reality, however, only a small number of shark species are potentially harmful. According to the ISAF, the three most dangerous species of sharks are

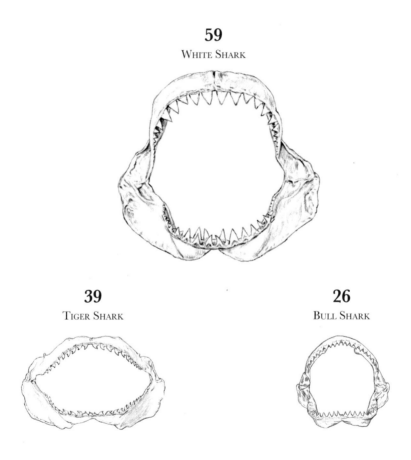

59
WHITE SHARK

39
TIGER SHARK

26
BULL SHARK

SHARK FATALITIES WORLDWIDE – THE "BIG THREE"
The "Big Three" species of sharks that are most commonly involved in incidents with humans, with their estimated human deaths.

the White Shark (*Carcharodon carcharias*), Tiger Shark (*Galeocerdo cuvier*), and Bull Shark (*Carcharias leucas*). Indeed, as this book went to press, the "Big Three" had been responsible for 488 unprovoked attacks on humans since 1958, of which 124 people tragically lost their lives. Oceanic Whitetips (*Carcharhinus longimanus*) might even displace one of these big three species, since many victims of maritime disasters are believed to have been killed by this species. Although the likelihood of being bitten by a shark is low, it is useful to remember that this happens only when we enter their world, which in many senses is a foreign wilderness to humans.

Mistaken identity

So why do sharks bite people? Are sharks mindless eating machines, attacking any animals below them on the food chain? Many people inaccurately assume that sharks are indiscriminate eaters with insatiable appetites. As we discussed in Chapter 7 (see page 87), some species of sharks are generalists and not too choosy about what they sink their teeth into. Sharks that live far offshore in food-poor environments, for instance, must be less discerning in order to survive. When these sharks encounter something that could provide nourishment, it is irrevocably baked into their biology to conclude that it is food—eat first, ask questions later. How else could one explain the stomach contents of White Sharks, Tiger Sharks, and Blue Sharks (*Prionace glauca*), which include tin cans, suits of armor, wine bottles, car tires, boots, and so on? Hardly nutritious fare. Still, at around 25 species, sharks of the open ocean surface waters represent a minority, and most occupy habitats in which we will very, very rarely encounter them. However, the remaining 500-plus species of sharks have very diverse diets and many are quite specialized in what they eat. Like any other animal, sharks also eat only as much as is required to meet their metabolic requirements and do not wantonly kill for the heck of it. Nor will they feed if their stomachs are already full. The perception that sharks have insatiable appetites is purely myth.

While we're at it, let's also vanquish the most pervasive myth about sharks. An overabundance of evidence conclusively shows that no shark includes humans as a regular part of its diet and no shark specializes on human beings.

In the vast majority of human-shark incidents witnessed by onlookers or described by the victim themselves there is a single bite or sometimes two, after which the shark disengages and swims away. The person is not consumed by the shark. If they see us as prey, why would they do that? The answer is simple: most sharks do not like to eat people and none do so regularly. Many sharks, even those who are generalist foragers, will reject human flesh almost instantaneously. Despite possessing exquisite senses, Blacktip Sharks in murky waters will apparently sometimes confuse human hands and feet with the small fish they eat and mistakenly bite. Almost instantaneously, taste receptors lining the inside of the mouth recognize the error and the shark releases its grip. Moreover, all of the sharks considered the most dangerous don't necessarily bite more frequently than other sharks. However, an error or exploration by a White Shark or Tiger Shark (see below) is more likely to have more extreme consequences than one from a blacktip.

Moreover, as packets of complete nutrition for sharks that prey on marine mammals, like White Sharks, humans fail again. Our bodies lack caloric density compared to the nutritious blubber of seals and whales, so we are simply not a desirable food source. In fact, it benefits them to not waste their time consuming a human buffet after tasting our flesh, since it is such a low-value meal. That may sound like a very cold description of such a horrific event, but this instinct may have saved the lives of many victims in the past, although we note with sadness that some people do not survive these exploratory first foraging events.

Not all shark bites can be simply attributed to mistaken identity, however. Many shark bites may be caused by a shark investigating a human being as potential prey. Once they have determined that we are unpalatable, they leave. Attacks on humans, for example, by juvenile White Sharks, might happen when sharks are learning how to hunt, especially since they might lack the visual acuity to distinguish pinnipeds (seals) from swimmers and surfers. It is unknown how often mistaken identity bites occur in the lives of sharks, but the evolutionary value of taking a test bite seems clear: to prevent swallowing prey that may be poisonous, of low nutritive value, too hard, or otherwise potentially harmful to the shark.

NEVER THE TWAIN
SHALL MEET

In areas where the potential for being harmed by a shark exists, measures are often implemented to protect beachgoers. In theory, separating people from sharks would seem to be a great idea, but the devil is in the details. The most commonly used barrier is a mesh net, chosen primarily because these are cheap and assumed to be effective. Mesh nets date back to 1907, when a steel structure was erected to keep the sharks away from a beach in Durban, South Africa. Although a variety of other materials have been tested for use as shark barriers, mesh beach nets are used most often because they are cheap and successful. In fact, they are too effective because sharks, along with sea birds and other marine life, can entangle themselves in the mesh and die. So high is the rate of mortal entanglement that some have referred to these mesh nets as fishing devices or even "walls of death" more so than barriers. Sadly, most of the sharks that die in these nests are harmless and some are protected threatened species. It could also be argued that beach nets are designed more to allay the fears of swimmers and to protect the economic interests of coastal communities than to separate them from sharks. Yet in reality these barriers are actually quite leaky in some cases and provide only a false sense of security.

Newer nets are constructed of more rigid plastics and thus are more durable, have reduced operating costs, and ensnare fewer or even no sharks and other marine life. In some scenarios nets have been replaced by drumlines, which are basically baited hooks at fixed locations attached to floats. The idea is to lure in and hook sharks, so they do not approach too closely to a beach where people are swimming. Drumlines may be effective in protecting swimmers, but they are often still lethal to sharks.

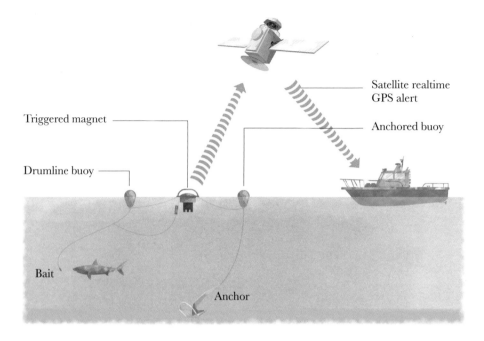

SMART DRUMLINES
*Innovative technology of the SMART drumline allows potentially
dangerous sharks to be caught and relocated.*

Welcome to the new age

Alternative devices in development include the Shark-Management-Alert-in-Real-Time (SMART) drumline. An upgraded version of a traditional drumline setup, this gadget includes an advanced solar-powered buoy with built-in GPS capabilities, which can trigger an alarm that is sent out to fishers and scientific researchers via satellite whenever a shark is hooked on a line. This allows the shark to be removed quickly from the fishing gear and relocated away from popular beaches, without killing it. Trials of this equipment in Australia have seen very encouraging results.

Another system being evaluated is the SharkSafe Barrier, which mimics the shape of kelp forests and physically blocks the entrance of large, potentially dangerous sharks. Containing electromagnets, these underwater fences are also designed as a repellant, unpleasantly overloading a shark's electrosensory system and scaring them off.

Finally, there are personal electric or magnetic shark deterrents in the form of bracelets, anklets, and surfboard attachments. As a model for a business, personal shark repellents for surfers and swimmers make an ideal product, a solution for which there is no problem, so to speak, since shark bites are extremely rare events in most locations. At the same time, however, these repellents may reduce anxiety among users. Rigorous testing has determined that some devices do significantly reduce shark bites on baits when they are active, but others only repel sharks when they get into very close proximity, less than 3.3 ft (1 m). Maybe a little too close for comfort!

A remarkably simple but effective way to prevent shark bites is to look for species that might be a threat and warn ocean users of their presence. The world-renowned flagship program for shark surveillance comes from the Shark Spotters of South Africa, who have developed a very reliable patrol system. Using high ground to make sightings easier, the Shark Spotters track sharks and monitor

SHARKSAFE BARRIER
Modern electromagnetic and physical barriers are being designed to exclude sharks from bather beaches.

their behavior near bather beaches. They then deploy color-coded flags to communicate risk levels and sound alarms to draw people out of the water when a shark is nearby. Not only has this system been incredibly successful for reducing shark attacks, it has also massively increased public trust and awareness.

In the past, planes and helicopters were used for shark spotting, but this was a time-consuming and expensive option. However, with the development of drones, aerial surveillance systems are having a resurgence. Drones have now become an invaluable tool in shark safety and scientific research, by providing shark surveillance programs with a much wider reach, even using image recognition algorithms that can monitor footage automatically as it is being streamed live from the drone via satellite. Alerts can even be synced with smart watches or phones, so that they can be received anywhere.

A sea change

Technologies to protect both people and sharks are being developed largely in response to outcries from conservation biologists and concerned members of the public, who are becoming increasingly aware of the myth that sharks pose a significant threat to us. The pendulum is swinging, albeit slowly, from fearing sharks to respecting them, and from the wrongheaded thinking that an ocean without sharks is a safer ocean to appreciating the essential ecological roles that sharks play as predators. The public is becoming increasingly aware of the serious declines in shark populations globally and the trophic cascades this could cause in marine ecosystems. People are starting to get savvy, and the media may be losing its stranglehold of misrepresenting sharks as scary monsters. So, what might this plot twist mean for the sharks of the future?

10

THE FINAL
CHAPTER?

IN HOT WATER

In the past, people probably imagined sharks to be indestructible: ultimate predators with an unequaled suite of adaptations and an amazing ability to survive in the cold, dark, mysterious depths of the oceans. How could such a beast—plentiful, swift, toothy, and elusive—be vulnerable to anything? Some people still harbor this belief today, but sadly, this could not be further from the truth. Despite existing for hundreds of millions of years, surviving multiple mass extinction events, glaciations, meteor strikes, wars, and nuclear disasters, sharks are now in very serious trouble. Their stories have changed dramatically thanks to human beings.

Imagine if we could compress planet Earth's entire history into one 24-hour period—from when the planet formed from dust in the void of space some 4.6 billion years ago, through all the movements of the landmasses, the formation of the oceans, and the production of the atmosphere, and through multiple ice ages … from the dawn of life some 3.7 billion years ago, to the rise of the dinosaurs … the fall of the dinosaurs … and the rise of mammals … all the way to today. If we were to scale all of our planet's history into one day, the appearance of human beings perhaps 400,000 years ago would equate to us arriving at the party at only about two minutes to midnight. We have not been around long at all in the Earth's grand scheme. Yet in that time we have caused unprecedented damage to our planet. The question is: Are sharks capable of surviving the human-dominated era (dubbed the Anthropocene) and the sixth great extinction we are now causing?

On average, shark populations have declined by as much as 71 percent globally over the last century. As many as a third of all elasmobranchs are threatened and in excess of 300 different species of chondrichthyan fishes are at risk of extinction. More species are added to the IUCN Red List of Threatened Species every single year, including the iconic Scalloped and Great Hammerheads (*Sphyrna lewini* and *S. mokarran*, respectively), which are now flagged as Critically Endangered; the stunning Whale Shark (*Rhinocodon typus*) and

SANDBAR SHARK
IUCN threat category: Endangered
Population trend: Decreasing
Threats: Overfishing; residential and commercial development

WHITE SHARK
IUCN threat category: Vulnerable
Population trend: Decreasing
Threat: Overfishing

SPINY DOGFISH
IUCN threat category: Vulnerable
Population trend: Decreasing
Threat: Overfishing

SMALLSPOTTED CATSHARK
IUCN threat category: Least concern
Population trend: Stable
Threat: Overfishing

THREATS TO OUR FEATURED SHARKS
*The IUCN has flagged three of our four featured species as threatened,
with overfishing being the main current threat to them all.*

Basking Shark (*Cetorhinus maximus*), which are both classified as Endangered; and even the magnificent White Shark (*Carcharodon carcharias*), which is now Vulnerable to extinction in the wild. Their dramatic population declines over the last few decades have led to sharks and their relatives being considered one of the most threatened groups of animals on the planet today.

Sharks are especially vulnerable to population declines because of their conservative, or slow, life history strategies—the one feature shared by every single species of shark and ray. As you have learned throughout this book, sharks are typically very slow-growing with a long life span, are late to reach maturity, and invest significant time and energy in reproduction. These are life history characteristics more in common with an African elephant than with an albacore tuna or barracuda. This means that sharks have a poor rebound potential compared to bony fishes, which have faster life history characteristics (for example, faster growth, shorter generation times, and higher fertility). This strategy worked wonderfully for sharks before we came along, as it had a natural balancing effect in their ecosystems; shark populations lived within their means, so to speak, and their numbers remained relatively stable (except during major extinction events). So, the ecosystems in which sharks swam were in harmony. However, in the face of persistent, major, and unprecedented human impacts, this life history strategy is failing, as sharks simply cannot recover rapidly when their populations are depleted. Bony fishes and other organisms with faster life history characteristics have the potential to bounce back when their stocks are overfished, but not sharks.

Estimates of shark mortality caused by human fisheries—the biggest current threat to sharks—are difficult to make with confidence, but one published source estimated that about 100 million were killed annually between 2000 and 2010. And far from slowing down, an assessment published in 2024 determined that global shark mortalities continue to rise, by between 4 and 7 percent every year. That means, if you have an average reading speed, by the time you have finished this chapter, as many as 3,700 sharks will have died at human hands globally.

GONE FISHING

The list of threats that sharks face is long and daunting, but without a doubt the greatest and most serious current danger is posed by fisheries. Overfishing of sharks and batoids, resulting from both incidental and targeted catch, is a much greater danger than habitat degradation, pollution, climate change, and so on (at least until these competing threats overtake it). Among shark fisheries, many people assume that the biggest driver of population declines is shark finning. Shark fin soup is a Chinese delicacy, traditionally served to guests as a symbol of status or for rites of passage since the Song Dynasty (960–1279 BCE). After normalization between the United States and China began in the 1970s and the Chinese ban on the dish was lifted in 1987, the demand for fins—and thus fisheries for the sharks to which the fins were attached—exploded, and ever since the most economically valuable part of most sharks has been their fins. Shark fin soup is now a part of the culture in East and Southeast Asia, including China, Hong Kong, Taiwan, Singapore, Malaysia, Vietnam, and Thailand.

However, it is important to note that the term "finning" expressly refers only to the disgracefully wasteful practice of removing the fins at sea and discarding the lower-value carcass. Fins generally can be legitimately traded when they are taken from sharks which have been legally landed and returned to port with their fins attached. While finning has been banned at many levels around the globe, it continues in many countries thanks to absent or poor regulations and/or enforcement. In the US, trade in shark fins was banned as part of a larger spending bill, a move popular with conservationists in general but less so with many shark biologists, who think it will damage the already well-managed US shark fishery. This in turn may encourage increased illegal finning elsewhere to meet global demands. Today, finning is still a significant problem that affects some of the most vulnerable species, and thus it merits a place of prominence in the list of existential threats to sharks. However, it is not the only type of fishery that impacts shark populations.

Sharks have been fished for human use for thousands of years. Sharkskin was once used for making tools; the ancient Greeks used it as sandpaper; and it improved the grip on sword hilts in China and Japan, as far back as the 2nd century. In the 18th century, shark leather, known as shagreen, became especially popular for clothes and accessories, such as glasses cases, luggage, and jewelry boxes. In the 19th century, shark liver oil, especially from Spiny Dogfish (*Squalus acanthias*), was used for lighting, tanning, waterproofing, and as a lubricant in tools and mills. It was also a very important source of vitamin A and often used as an alternative to cod liver oil as a dietary supplement. Biomedical uses for sharks include skin grafts, medical creams, and therapy for arthritis (but not cancer, despite some persistent claims about shark cartilage). Squalene from shark livers is also an ingredient in a range of beauty products and some vaccinations. Many communities globally continue to rely on small-scale artisanal shark fishing for food, products, and income. Fresh and preserved shark meat is eaten in developing coastal and island countries, and shark steaks can also be found in seafood markets.

The major problem with shark fisheries is one of scale. Up until the mid- to late 20th century, sharks were fished in a relatively sustainable manner by subsistence or artisanal fishers using primitive, low-tech methods like small nets or hook-and-line. Modern fishing fleets are a different beast entirely. Today, fishers can extract sharks from our oceans at such an unprecedented rate that they simply cannot recover. As a group, shark and batoid fisheries represent less than 1 percent by weight of total marine capture fisheries. But one thing is clear: the estimated 777,000–915,000 tons caught annually across the globe is simply unsustainable and a major driver of declines in every species of those groups of batoids and sharks threatened with extinction.

Overfishing of sharks refers to both the targeted capture of sharks and also their bycatch—incidental or unwanted catch in fisheries targeting other commercially valuable species. Both represent challenges for the management and conservation of sharks. Most fishing gear, including longlines, trawls, purse seines, and gillnets incidentally catch sharks to some degree and the IUCN attributes bycatch as a significant threat to 66.9 percent of all species of sharks threatened with extinction. Mortality rates for sharks caught as bycatch vary but are high for many

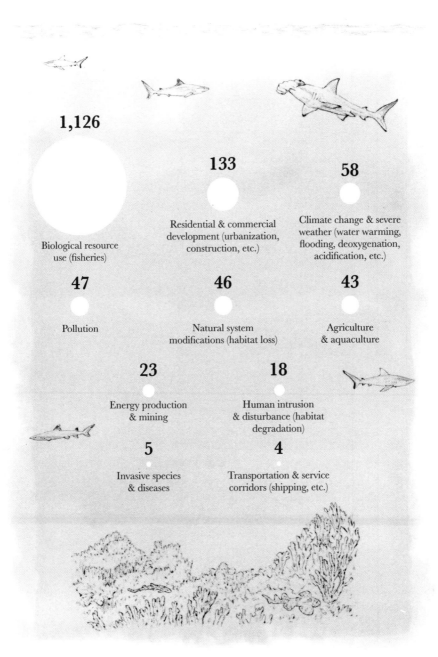

1,126

Biological resource
use (fisheries)

133

Residential & commercial
development (urbanization,
construction, etc.)

58

Climate change & severe
weather (water warming,
flooding, deoxygenation,
acidification, etc.)

47

Pollution

46

Natural system
modifications (habitat loss)

43

Agriculture
& aquaculture

23

Energy production
& mining

18

Human intrusion
& disturbance (habitat
degradation)

5

Invasive species
& diseases

4

Transportation & service
corridors (shipping, etc.)

THREATS TO SHARKS
*While there are many threats to sharks, by far the greatest
and most pervasive, at least currently, is overfishing.
The above chart estimates the number of species currently at risk.*

species. For example, more than 95 percent of Cuban Dogfish (*Squalus cubensis*) caught on bottom longlines may be released alive, but about half will die soon after. Great Hammerheads are particularly at risk of succumbing to stress after being caught and released. In some cases, indelicately removing the hook from the shark's mouth breaks the lower jaw. Moreover, even bycatch that consists of few or no sharks may include the forage base (food source) for sharks, and thus may have indirect negative consequences for sharks or for the ecosystems in which they reside.

Many sharks are also killed by ghost fishing, when fishing gear that is lost or discarded continues to indiscriminately entangle sharks and other marine organisms. This problem has been exacerbated by the transition from nets and ropes that degrade over time to more durable plastics, which can remain intact for decades. Sharks are particularly vulnerable because they often bite objects in the water to investigate them, or attempt to eat other entangled animals, and this can lead to entanglement in ghost gear. Curiosity kills the catshark, so to speak.

All for one and one for all

When we discuss how to conserve shark populations, it can become very easy to point the finger of blame at others, such as East and Southeast Asia, for their disproportionate impact on shark populations as a result of the demand for fins. But this is an oversimplification. Protecting shark stocks and aiding their recoveries are complex and dynamic problems.

It might surprise you to learn that India, Indonesia, Mexico, Spain, and Taiwan consistently have the highest landings of sharks every year. Additionally, many other countries are harvesting sharks in unsustainable numbers, including China, Argentina, Pakistan, Malaysia, Japan, Australia, Brazil, Canada, Libya, Mozambique, Russia, South Africa, the United Kingdom, and the United States. While fins remain the most profitable part of the shark, the demand for shark meat far exceeds demand for fins. Shark meat is widely consumed, especially in Brazil, India, Italy, Japan, South Korea, the US, and most of Southeast Asia. If you have ever had fish and chips dubbed "huss," "flake," or "rock salmon," in either Australia or the UK, you have most likely eaten shark meat. Shark overexploitation is a global issue that affects us all.

HOME SWEET HOME

Habitat degradation is another major issue affecting sharks, and one that would be the focus of our concern were it not for the overwhelming impact of overfishing. Habitats may be physically damaged or lost completely because of dredging, shipping, pollution, urbanization, and ocean acidification and deoxygenation. They can also be degraded by the loss of prey species on which sharks depend. Habitat degradation is a particular problem for sharks inhabiting tropical and temperate coasts, due to their close proximity to human activity. One vivid example is the impact that the development of a large resort complex on North Bimini, in the Bahamas, has had on Lemon Sharks (*Negaprion brevirostris*), which had utilized parts of the impacted area as a major nursery. Dredging and removal of mangrove habitat led to a decreased survival rate of juvenile Lemon Sharks in the system and the coverage of seagrass (habitat for their prey) also declined.

Chemical burn

Habitats can also be degraded over varying time scales by different forms of pollution. Oil spills can be almost immediately lethal to filter-feeders like Whale Sharks, as crude oil can clog their gill rakers, making it impossible for them to breathe or eat. What's more, oil also degrades into very toxic chemicals. Studies performed in the Gulf of Mexico following the Deepwater Horizon oil rig disaster in 2010 discovered Silky Sharks (*Carcharhinus falciformis*) with significantly higher levels of polycyclic aromatic hydrocarbons (PAHs) than normal, and multiple other species of sharks also had elevated polychlorinated biphenyl (PCB) levels. These toxins can cause neurological impairment and organ damage, and a host of other impacts such as slower growth rates, reduced fertility, and/or weakened immune systems.

Another insidious and potentially huge problem is litter. As much as 1.8–10.0 percent of the 660 billion lb (272 billion kg) of plastics produced globally finds its way into the ocean every year. There is no denying that we live in the

DESTRUCTION OF NURSERY HABITATS
*Clearing and coastal urbanization are destroying mangrove nursery habitats
in the tropics, reducing the recruitment of Lemon Sharks to the adult population.*

Age of Plastic. When we removed the plastic packaging ensnaring a Sandbar Shark (*Carcharhinus plumbeus*) around its gills in 2016, we considered the incident anecdotal and rare, but such reports are now commonplace. In one incident, for example, scientists discovered a Shortfin Mako (*Isurus oxyrinchus*) so entangled in plastic debris that its growth was stunted and its spine deformed.

Plastics break down into smaller fragments, after which they pose a different suite of problems. Pieces can be ingested by filter-feeders like manta rays, and the buildup of plastic fragments in their guts has been implicated in the deaths of several Whale Sharks that have washed up across Southeast Asia and South America in recent years. Microplastics have even been found in the guts of deepwater sharks that live far offshore. Predatory sharks can also indirectly ingest microplastics which have accumulated in the tissues of their prey. As we write, there is still uncertainty about the toxicity of microplastics and their overall impacts on the health of sharks.

Keep the noise down

It's not only chemical pollutants that can affect sharks. In recent years, scientists have also become increasingly aware of the impacts of noise pollution on marine life. Sound travels faster and farther in water than air, and we expose our oceans to an incessant cacophony of construction, underwater drilling and pile-driving, motor noise from shipping lanes and recreational boating, sonar, seismic surveys, and military and fisheries explosives. Scientists in Australia have learned that White Sharks and many coastal species of sharks, such as Sicklefin Lemon Sharks (*Negaprion acutidens*), Bronze Whalers (*Carcharhinus brachyurus*), Grey Reef Sharks (*C. amblyrhynchos*), Dusky Sharks (*C. obscurus*), Sandbar Sharks, Scalloped Hammerheads, and Zebra Sharks (*Stegostoma fasciatum*), are affected by noise pollution to varying degrees. Artificial sounds in their environment can elicit startle responses and affect how the sharks forage for food. In the long term this could shift their natural ranges, constrict their available habitats, and seriously impact their fitness.

CLIMATE CHANGE

Another new threat facing sharks—perhaps the parent of all threats—is climate change. There is no doubt that the climate crisis will remake the natural world, and it is very likely that every single ecosystem and nearly every species will be adversely affected in unimaginable and unpredictable ways. We are only just beginning to understand how climate change will affect sharks.

By the year 2100, average sea surface temperatures are projected to rise by at least 1.8–2.0°F (0.5–0.6°C), and they may not necessarily stop there. These higher temperatures can seriously affect many ecological processes (such as nutrient cycling), which can alter community structure, food availability, and predator–prey dynamics. As we all know, rising temperatures also massively affect weather and climate. This may cause changes in precipitation patterns—that is, deluges as a result of the increased intensity and frequency of tropical

BLEACHED CORAL REEFS
Reef habitats are declining due to bleaching events associated with climate change.

storms—and droughts as well. The subsequent changes in salinity will affect the abundance, biodiversity, and distribution of sharks, both nearshore and in oceanic systems. Additionally, this could lead to large-scale changes to oceanic circulation patterns, such as the slowing of the Gulf Stream. Tropical marine ecosystems, especially coral reefs, are particularly vulnerable to these impacts, and the degradation or loss of coral reefs will reduce the availability of shark prey and suitable habitat for many sharks.

Climate change could also drive the loss of critical nursery habitats, as rising sea levels and changing precipitation patterns will change important coastal habitats like mangrove forests and estuaries This could wipe out vital nurseries and seriously hinder shark reproduction. We are already seeing the first signs of this in Winyah Bay, South Carolina, an estuary that serves as a major nursery for juvenile Sandbar Sharks. There, increased precipitation has led to decreased salinity and thus changes in the biological community sensitive to these changes, including shark populations.

Large-scale changes to oceanic chemistry, specifically increased acidity and decreased dissolved oxygen, are also a consequence of climate change. From laboratory studies, scientists have discovered that these conditions can lead to serious skin damage in Puffadder Shysharks (*Haploblepharus edwardsii*) and retard egg development to such a serious degree that newborn Bamboo Sharks (*Chiloscyllium punctatum*) cannot survive. In the wild, changing conditions have already started to slow the growth of Port Jackson Sharks (*Heterodontus portusjacksoni*) in South Australia, delaying their development to maturity. Hypoxic zones (waters with low oxygen) are already expanding, causing the available habitats for Blue Sharks (*Prionace glauca*) to seriously contract, shifting their distribution in the Atlantic Ocean. Changes in behaviors have also been observed. Port Jackson Sharks and Smooth Dogfish (*Mustelus canis*) hunt less efficiently under these altered conditions, while Lesser Spotted Catsharks (*Scyliorhinus canicula*) swim differently in warm, acidic waters.

Sharks as a group are already facing massive existential threats from a myriad of different hazards, but scientists, policymakers, and fishery managers are beginning to enact effective measures to help some stocks recover and prevent further declines in others. Climate change has the real potential to nullify all of these. Our survival as a species requires a healthy biosphere, and one component of that is marine ecosystems with the appropriate biodiversity of sharks—that is, the mix of shark species at population sizes that evolution has established as optimal. Every molecule of carbon dioxide that we can keep from entering our atmosphere and every fraction of a degree in temperature we can claw back now will be beneficial to sharks of the future, and indeed all of us.

PLOT TWIST

Thankfully, there are hopeful signs that, when it comes to shark conservation, hearts, minds, and actions are changing in the right direction. People still fear sharks, but less so than in the past; and awe, respect, and appreciation for the irreplaceable roles that sharks play are on the rise. As a result, shark ecotourism is mushrooming, in part because sharks are so captivating, and yes, in part because of the adrenaline rush many get from having so powerful and potentially dangerous a beast swim by within arm's reach. But shark ecotourism is more than just a thrill; it also has incredible power to shift public attitudes, as it provides an excellent platform for the education of tourists and the local community alike. Studies have shown that interactions with animals in the wild can elicit a serious emotional response so powerful that it makes people more conservation-minded, and seeing sharks in their habitat peacefully being sharks supplants the violent depictions that bombard us in the media. As this can play such an important role in countering negative perceptions, ecotourism is now becoming a significant driving force for the conservation of marine ecosystems.

Shark tourism can bolster local economies and can switch the perceived value of sharks. For instance, in South Africa, US $6,074,000 was spent at White Shark cage-diving attractions in 2011 compared to the measly $478,000 generated by the country's shark fisheries. In French Polynesia, individual sharks are estimated to be worth a whopping US $545 per lb ($1,200 per kg) as an ecotourism resource, compared to only $0.69 per lb ($1.5 per kg) when fished for their meat. The continuous nature of this type of business means that live animals have much more value than they would if they were fished, thus incentivizing local people to protect their sharks. These benefits often accrue to those capable of exploiting the economic opportunities tourism offers, and not necessarily to the broader local community. Moreover, they are not without costs to sharks —attracting sharks to baited stations may impact the their behavior, nutrition, and/or predator–prey dynamics to some degree. Even so, the lesson is clear: sharks are more valuable alive than dead.

A never-ending story

In our experience, there is great hope for sharks within the scientific community. We know that many shark populations are not threatened in the wild—the Lesser Spotted Catshark, for instance—and for those that have been depleted, the data demonstrates that stocks can recover if they are managed effectively. One study found that four of the five shark stocks assessed would begin to increase within ten years if protective measures were introduced today. Populations can rebound given the chance.

SHARK DIVING ECOTOURISM
*Snorkeling, SCUBA, and cage diving with sharks helps raise awareness,
creates platforms for education and engagement, and generates revenue,
which all can aid shark conservation.*

In fact, in some regions recoveries have already started. After cataclysmic decreases in populations in the Atlantic from the mid-1970s, there is strong evidence that shark stocks have now started to rebound after the implementation of fisheries management plans in the United States. These include coastal populations of Sandbar Sharks, Blacknose Sharks (*Carcharhinus acronotus*), Tiger Sharks (*Galeocerdo cuvier*), and Bonnetheads (*Sphyrna tiburo*). Spinner Sharks (*Carcharhinus brevipinna*) have increased in abundance by as much as 14 percent. What's more, these measures are also reversing anthropogenic size selection, in which sharks had been decreasing in average size for several decades because of selective fishing. This gives us undeniable proof that if we implement strict, cautious, science-based management plans for sharks, they can slowly recover, even after serious declines.

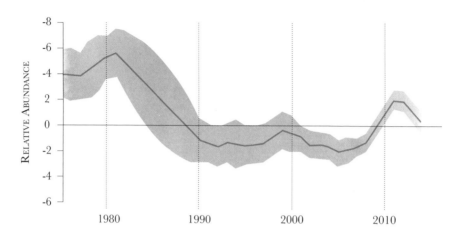

SANDBAR POPULATIONS
*Since the implementation of fisheries management plans in the US
in 1993, populations of several species, including Sandbar Sharks,
have started to rebound.*

Making the change

Given that currently the top threat to sharks is overfishing, the first step to conserving sharks is to make fisheries more sustainable. From a practical perspective, a total ban on shark fishing is not realistic. Many communities around the
world rely on sharks for subsistence or income, and past bans have proven
ineffective because enforcement was too expensive. It is therefore preferable to
work with, rather than against, shark fishing interests to devise management
strategies which, where possible, protect both sharks and local stakeholders,
including the fishers.

Most shark scientists believe that sustainable fisheries are pivotal to shark
conservation. Several countries around the world have already reported sustainable fisheries for sharks, including the United States and Australia. For
example, even though they are considered Vulnerable by the IUCN, the Pacific and Atlantic US fisheries for Spiny Dogfish have been certified as sustainable, and monitoring of the stock has shown that their populations are stable,
despite being fished. Many factors must be considered to make shark fisheries

sustainable, however, and there is no one-size-fits-all strategy because sharks are so diverse. The first step should always be to ensure that all management plans are based on rigorous scientific assessment of the stock and are tailored to the species in question. Achieving sustainable fisheries also requires protection not only of the target species (the sharks themselves) but also the forage base (their food sources). Protective measures must then also be continuously reevaluated to determine if the population is remaining stable.

One effective strategy is to implement strict quotas (that is, a limit on landings) for shark fisheries. Another effective management action is time-area closures in which shark fishing is prohibited in critical habitats at specific times—nursery areas during breeding season, for example. It goes without saying that Endangered species of sharks should not be fished. Another effective conservation measure is no-take zones, where sharks cannot be fished under any circumstances. This has led to the designation of Marine Protected Areas (MPAs) by many countries. In these regions sharks are protected from fisheries to varying degrees, but each MPA is governed by its own unique policies. Shark fishing may be banned entirely, restricted to specific areas, or allowed only for small-scale artisanal vessels or local residents. Since many MPAs protect entire ecosystems, the benefits extend to most or all of the residents, not only iconic or charismatic species.

Several nations—the first of which was Palau, in the western Pacific, in 2009—have designated their entire Exclusive Economic Zone as shark sanctuaries, where commercial shark fishing, finning, and sale of any shark products are banned. Shark abundance and species richness (how many different types of sharks there are) are already rising within some of these shark sanctuaries, especially in French Polynesia, Palau, and the Maldives. The recovery process is very slow, but there is hope that these safe havens might offer a stronghold, where sharks can boost dwindling populations and potentially even restock themselves in the areas outside of the sanctuaries over time.

The type of fishing gear used can also have an impact on shark catch rates and mortality. Altering mesh shape and net sizes can reduce shark and ray bycatch in trawl fisheries. Gillnets (resembling volleyball nets), which are designed to entangle their catch, are generally the most destructive to sharks.

Up to 70 percent of Gummy Sharks (*Mustelus antarcticus*) caught in gillnets die, compared to only 8 percent caught on longlines, which employ baited hooks. This has resulted in the United States banning the use of large-mesh drift gill-nets for swordfish 3–200 miles (5–320 km) offshore as of 2023. In 2020, the nation of Belize, in South America, took the ground-breaking action of entire-ly banning the use of any gillnets in their waters. Rather than exclusively crim-inalizing the use of gillnets, the Belizean government instituted a buy-back program in which fishers who relinquished their gillnets were compensated for the costs associated with purchasing alternative gear. Maybe other nations will be inspired to follow suit.

The technological age

In our technological world, improving equipment and developing gadgets that can make shark fisheries more sustainable may play major roles in combating overfishing. This can be as simple as changing hook shapes and sizes, which have been shown to have a huge impact on shark post-release mortality. More advanced methods involve attempting to repel sharks from gear targeting other species, by employing magnetic hooks or those made with rare-earth metals. However, as sharks can become habituated to these hooks or the metals degrade rapidly, in their current form they may be effective only for a short time. But engineers continue to work on it.

A recent development is the SharkGuard, a device which produces a power-ful, short-range electric pulse that is thought to overstimulate the sharks' elec-trosensory organs and cause sharks to flee from fishing gear. In testing the gad-get in longline tuna fisheries, initial trials have been very promising; with catches of Blue Sharks reduced by 91.3 percent. The technology is not yet perfect, as there is concern that the electrical fields may have slightly reduced tuna catches as well, which would obviously make fishers reluctant to use them in their work. Also, while the cost is decreasing, this technology is currently very expensive and difficult to implement, which may make it prohibitive for some fishers at the moment. But if we could develop an inexpensive, effective SharkGuard-like device, it could reduce shark and ray bycatch around the world. In one study, inexpensive lights on gillnets significantly reduced elasmo-branch (and turtle) bycatch, which made hauling the net easier, and didn't reduce the target catch.

Two minutes to midnight

There is no denying that the scale of shark declines is huge and the solutions to turn the tide are complex, expensive, dynamic, and endlessly complicated. But scientists and conservationists believe there is hope for sharks, and that we have the conservation tools and know-how to help them recover. Management plans can work if they are intelligently implemented and enforced. But the problem is undeniably urgent and we absolutely must make changes, now. There is no time to waste.

However, in the face of this seemingly insurmountable challenge—figuratively at the base of a mountain that looks enormous and unsurpassable—we must remember that human beings are endlessly creative and capable of achieving seemingly impossible goals. We can design our habitats, generate artificial intelligences, perform organ transplants, and construct an electronic computer network that allows people to communicate with each other from all over the world in the blink of an eye. If we can muster the will to focus and turn all this human ingenuity toward saving the planet, rather than exploiting it, there is no telling what we might achieve. A world without sharks is simply too horrifying to allow.

LIST OF SPECIES

African Dwarf Sawshark (*Pristiophorus nancyae*)

American Pocket Shark (*Mollisquama mississippiensis*)

Angel Shark (*Squatina squatina*)

Atlantic Sharpnose Shark (*Rhizoprionodon terraenovae*)

Banded Wobbegong (*Orectolobus halei*)

Basking Shark (*Cetorhinus maximus*)

Bigeye Thresher (*Alopias superciliosus*)

Blackmouth Catshark (*Galeus melastomus*)

Blacknose Shark (*Carcharhinus acronotus*)

Blacktip Reef Shark (*Carcharhinus melanopterus*)

Blacktip Shark (*Carcharhinus limbatus*)

Blue Shark (*Prionace glauca*)

Bluntnose Sixgill Shark (*Hexanchus griseus*)

Bonnethead (*Sphyrna tiburo*)

Bramble Shark (*Echinorhinus brucus*)

Broadnose Sevengill Shark (*Notorynchus cepedianus*)

Bronze Whaler (*Carcharhinus brachyurus*)

Brownbanded Bamboo Shark (*Chiloscyllium punctatum*)

Bull Shark (*Carcharhinus leucas*)

Caribbean Reef Shark (*Carcharhinus perezi*)

Chain Catshark (*Scyliorhinus retifer*)

Common Thresher (*Alopias vulpinus*)

Cookiecutter Shark (*Isistius brasiliensis*)

Cuban Dogfish (*Squalus cubensis*)

Dusky Shark (*Carcharhinus obscurus*)

Dwarf Lanternshark (*Etmopterus perryi*)

Eagle Shark (*Aquilolamna milarcae*)

Epaulette Shark (*Hemiscyllium ocellatum*)

False Catshark (*Pseudotriakis microdon*)

Finetooth Shark (*Carcharhinus isodon*)

Frilled Shark (*Chlamydoselachus anguineus*)

Galapagos Shark (*Carcharhinus galapagensis*)

Ganges Shark (*Glyphis gangeticus*)

Goblin Shark (*Mitsukurina owstoni*)

Great Hammerhead (*Sphyrna mokarran*)

Greenland Shark (*Somniosus microcephalus*)

Grey Reef Shark (*Carcharhinus amblyrhynchos*)

Gulper Shark (*Centrophorus granulosus*)

Gummy Shark (*Mustelus antarcticus*)

Halmahera Epaulette Shark (*Hemiscyllium halmahera*)

Horn Shark (*Heterodontus francisci*)

Japanese Wobbegong (*Orectolobus japonicus*)

Kaja's Sixgill Sawshark (*Pliotrema kajae*)

Kitefin Shark (*Dalatias licha*)

Lanternsharks (*Etmopterus* species)

Lea's Angel Shark (*Squatina leae*)

Lemon Shark (*Negaprion brevirostris*)

Longfin Mako (*Isurus paucus*)

Longnose Sawshark (*Pristiophorus cirratus*)

Megalodon (*Otodus megalodon*)

Megamouth Shark (*Megachasma pelagios*)

Narrownose Smoothhound (*Mustelus schmitti*)

Northern River Shark (*Glyphis garricki*)

Nurse Shark (*Ginglymostoma cirratum*)

Oceanic Whitetip Shark (*Carcharhinus longimanus*)

Pacific Angel Shark (*Squatina californica*)

Pacific Sleeper Shark (*Somniosus pacificus*)

Painted Horn Shark (*Heterodontus marshallae*)

Pelagic Thresher (*Alopias pelagicus*)

Pigfaced Shark (*Oxynotus centrina*)

Port Jackson Shark (*Heterodontus portusjacksoni*)

Porbeagle (*Lamna nasus*)

Portuguese Dogfish (*Centroscymnus coelolepis*)

Puffadder Shyshark (*Haploblepharus edwardsii*)

Pyjama Shark (*Poroderma africanum*)

Pygmy Shark (*Euprotomicrus bispinatus*)

Redspotted Catshark (*Schroederichthys chilensis*)

Ridged-egg Catshark (*Apristurus ovicorrugatus*)

Salmon Shark (*Lamna ditropis*)

Sand Tiger (*Carcharias taurus*)

Sandbar Shark (*Carcharhinus plumbeus*)

Scalloped Hammerhead (*Sphyrna lewini*)

Seal Shark (*Dalatias licha*)

Shortfin Mako (*Isurus oxyrinchus*)

Silky Shark (*Carcharhinus falciformis*)

Silvertip Shark (*Carcharhinus albimarginatus*)

Smallspotted Catshark (*Scyliorhinus canicula*)

Smalltooth Sand Tiger (*Odontaspis ferox*)

Smalltooth Sawfish (*Pristis pectinata*)

Smooth Hammerhead (*Sphyrna zygaena*)

Smooth Dogfish (*Mustelus canis*)

Southern Sleeper Shark (*Somniosus antarcticus*)

Speartooth Shark (*Glyphis glyphis*)

Spinner Shark (*Carcharhinus brevipinna*)

Spiny Dogfish (*Squalus acanthias*)

Spiny Sharks (Acanthodians—extinct class of early sharklike ancestors, including *Doliodus problematicus*)

Starry Smoothhound (*Mustelus asterias*)

Swell Shark (*Cephaloscyllium ventriosum*)

Tasseled Wobbegong (*Eucrossorhinus dasypogon*)

Tawny Nurse Shark (*Nebrius ferrugineus*)

Tiger Shark (*Galeocerdo cuvier*)

Tope (*Galeorhinus galeus*)

Whale Shark (*Rhincodon typus*)

White Shark (*Carcharodon carcharias*)

Whitetip Reef Shark (*Triaenodon obesus*)

Whitespotted Bamboo Shark (*Chiloscyllium plagiosum*)

Winghead Shark (*Eusphyra blochii*)

Zebra Shark (*Stegostoma tigrinum*)

INDEX

ACKNOWLEDGMENTS

Dan and Sophie would both like to heartily thank the whole team at UniPress Books/Princeton University Press for so expertly sculpting our words into such a lovely book! We would especially like to acknowledge Richard Webb—all editors should be so responsive, courteous, and brilliant! We also thank Jason Hook, Slav Todorov, and Jenny Manstead of UniPress Books, and Robert Kirk of Princeton University Press. We know illustrator Adam Hook only through his wonderful art, but what wonderful art! The look and feel of Adam's illustrations really set this book apart from others in the field. We are grateful for the entire village that nourished and grew this book!

Daniel Abel. A writer, especially *this* writer, does not ply his craft solely, and thus I appreciatively acknowledge the numerous family members, friends, colleagues, baristas, and mentors who knowingly or not enabled me to write this book. Both space limitations and the honest fear of omitting someone prevent me from naming them here (I have done so in earlier works).

Sophie Maycock. I want to thank David Fitter and Louisa MacKenzie, who are living proof that great educators create inspiration that lasts a lifetime. I also want to acknowledge the huge impact that Lewis Warren had on my life—I live every day for you and to make you proud.

To my Dad, thank you for all your help and support throughout the years, and to Cillian, thanks for being the motivation I desperately needed. I am grateful to all my beloved friends, but I especially owe a debt to Anne Harvey-Lynch for pushing me to believe in myself, and my colleagues at the European Elasmobranch Association who taught me so much.

Last (but very far from least), I am so grateful to my greatest cheerleader, devoted therapist, and absolute rock, Rob Hale. I genuinely could not have done this without your endless support and I only hope I can be half as great as you believe me to be. Hopefully this book teaches you to stay away from the Oceanic Whitetips!